Theodore Modis

Science with Street Value

A Physicist's Wanderings off the Beaten Track

Theodore Modis

SCIENCE
WITH
STREET VALUE

A Physicist's Wanderings off the Beaten Track

Bibliografische Information der Deutschen Nationalbibliothek
Die Deutsche Nationalbibliothek verzeichnet diese Publikation in der Deutschen Nationalbibliografie; detaillierte bibliografische Daten sind im Internet über http://dnb.d-nb.de abrufbar.

Bibliographic information published by the Deutsche Nationalbibliothek
Die Deutsche Nationalbibliothek lists this publication in the Deutsche Nationalbibliografie; detailed bibliographic data are available in the Internet at http://dnb.d-nb.de.

ISBN-13: 978-3-8382-1447-4
© *ibidem*-Verlag, Stuttgart 2020
Alle Rechte vorbehalten

Printed in the EU

To Mihali Yannopoulos

This is a personal account of true events and real people,
but only public figures, well-known personalities,
and Nobel laureates appear with their real names.

CONTENTS

I

1 – Signs of Disenchantment

Concealing all evidence of a hidden agenda, Ted engaged Pascal, his physics colleague, in the following discussion. Planets far from the Sun, such as Jupiter and Pluto, move around the Sun slowly whereas planets close to the Sun, such as Mercury and Venus move faster. At the same time, distant planets receive less sunlight per square meter than closer planets. Doing the math, Ted had deduced that there was a strict relationship between light received and orbital speed: the more light a planet receives, the higher its speed.

"They behave like living organisms," Ted concluded. "They move according to the amount of heat they receive. Turn on the heat on a frying pan full of ants and see what happens."

Pascal smiled. "If you are trying to prove that planets are living beings, you are not there yet," he mused.

But Ted insisted. "You know the ideal gas law. Gas molecules in a container will agitate as the temperature rises."

"You mean their mean velocity will increase," Pascal objected to anthropomorphizing molecules with the word "agitate."

"Yes, but the frequency of their collisions will also increase resulting to a higher pressure. You know what my wife's shrink suggested to her recently when she told him we've been fighting a lot? 'Get a bigger apartment!' Increasing the volume decreases the temperature and/or the pressure, in other words, the number of collisions. Do you suppose her shrink knows the ideal gas law?" Ted asked ironically.

"I doubt very much that he ever studied the laws of physics," said Pascal and continued," he may, however, know it instinctively. Behind every law of physics there is a fundamental truth. This psychiatrist may have gut knowledge of this truth, which for gases is described by the ideal gas law."

Ted decided not to push Pascal further for now; after all Pascal was the group's leader and this kind of discussion could put Ted's scientific reputation in jeopardy. But Ted wondered how many of the numerous physics laws he had been taught in university might be subconsciously known to non-physicists, and what use of them they may be inadvertently making.

The discussion with Pascal was yet another incident of Ted venturing away from hard-core science. Throughout his life he had become intrigued by topics that science considers taboos, such as miracles, deities, and the

1

supernatural, or by philosophical issues that science simply is not interested in, such as the purpose of life.

"When was the first time I was 'unfaithful' to science?" he wondered. Stretching his memory as far back as he could go he pictured himself as a pre-teen boy with a book at his hands. Its title was *The Great Initiates* by Édouard Schuré.

The book was a study of the secret history of religions and discussed Rama, Krishna, Hermes, Moses, Orpheus, and Pythagoras. The discussion was beyond the little boy's comprehension and he did not read the entire book. But it intrigued him so that somehow this book followed him around the world and his first copy made its way into his library where it still sits today.

But this could not have been the beginning of his "misbehavior." After all, how correct a notion of science can a child have? He remembered another incident when he was a young teenager sitting at dinner table when his older sister, whom he admired for her wisdom and maturity of thought, uttered unexpectedly:

"I'll be for ever indebted to the person who will tell me what the purpose of life is." The question landed like a bomb and froze all conversation. In the absence of a father, Ted's mother felt called upon to take up the challenge. Ted beamed with anticipation.

"The purpose of life is to grow up, get married, and have children, who themselves will later do the same thing," she slowly articulated as she verbalized her thoughts. "What else can it be?" she added almost to herself.

"Well, if that's it, ..." his sister left her phrase unfinished visibly unsatisfied with the answer. Ted retained that the purpose of life couldn't be that simple.

But there was also that white night years later and a few days before Ted left home to begin his physics studies at Columbia University. Together with two other high-school friends he had stayed up all night drinking beer, smoking cigarettes, and pondering on life's existential questions. At some point during the early hours of the morning the topic turned to the existence of God. Ted stopped participating in the discussion claiming that his knowledge of physics was not good enough yet to tackle such questions. In his mind he had clearly linked God with physics.

It wasn't long, however, before his first taste of disenchantment. Two yeas later at Columbia they studied the atom. He found that atomic physics involved too many approximations; left questions unanswered, and employed kitchen-like recipes. This was a far cry from the clean-cut Euclidian approach to classical physics that had fascinated him in high school. There was more such disenchantment for him in the following years. Nevertheless he persevered with his studies as planned; he had been betrothed to science since early childhood and not becoming a physicist was not an option.

When he finally arrived at graduate school and began the research for his Ph.D., he joined a group of physicists preparing to carry out a large-scale experiment at Brookhaven National Laboratory. The group consisted of three post docs, a professor, Jack Steinberger (later Nobel laureate) and another student, Steve. The group was going to study the behavior of the neutral K–meson particle.

One day following a group meeting one of the postdocs challenged the two graduate students: "I'll give a quarter to whoever can write down the state of the neutrino particle that will be coming out from the decay of the K–meson particle we are trying to study." In ordinary language he would be asking: "What do we know about this neutrino particle, how big is it, what charge does it have, how fast is it moving, etc.?"

Ted and Steve scrambled to the blackboard where they worked collectively for about half an hour. An initially simple equation became increasingly complicated as more terms were added to it. At the end they stopped, unable to make further additions or corrections; the postdoc seemed satisfied. "You deserve the quarter," he said. "You can split it between the two of you!" At that moment Jack walked in, saw the equation on the board and remarked, "It is probably not far from the truth."

Ted had found it objectionable that the theoretical description turned out to be so complicated and that the "high priest" found it only probably close to the truth. Ted resented complication even in precisely described phenomena. When they studied the bicycle in a classical mechanics course two years earlier, the equations had proved equally complicated, if exactly accurate in that case. One was left marveling at the contrast of how simple it is to ride a bicycle and how complicated it is to describe the phenomenon scientifically. Complication reduces not only the elegance of a formulation but also its utility. A very complicated theoretical expression is generally of little use.

Elegance is associated with simplicity. Classical physics is excellent in describing billiard ball movements and making accurate predictions. The difficulty comes from putting many balls together and *many* for physicists can be anything greater than three. Molecules in a volume of gas behave very much like billiard balls, but there are too many of them and they bounce too often. However, thermodynamics, the branch of physics that studies gases, makes predictions by focusing only on the macroscopic global variables: temperature, pressure, and volume. The bottom-up approach, tracking individual molecules, taxed the ingenuity of the best minds in physics for at least one-hundred-and-fifty years and has served only in understanding, corroborating, and justifying the relations established experimentally between the overall variables.

By and large, complicated formulations in physics have remained academic exercises.

2 – There Cannot Be Magic

Ted was not a science purist but had been passionately attracted to physics from a tender age. He had always liked math and did well in high-school science courses. And yet, when people later asked him why he had become a physicist, he whimsically responded, "Because my mother had been overly impressed by the atomic bomb," which, like all jokes, contained a grain of truth.

He had not in fact become a typical physicist. In physics circles he often reasoned as a rebellious child. Once he asked Jack's opinion about an unconventional idea he had for an experiment. Ted respected Jack's opinion not the least because of Jack's Nobel Prize.

"What do you think, Jack, about doing an experiment to check the validity of statistics?"

"What do you mean?" asked Jack, puzzled.

"Well, we throw a die many times, thousands of times, and we monitor how often the six appears and how close this outcome comes to one sixth of the time. We know what deviations to expect from the statistical error. This error gets smaller and smaller as we increase the number of throws. We can make an experiment to check if the real outcome is always within the expected error, or whether there is a violation, if tiny, of this law of statistics observable only at very large numbers."

"But, Ted," Jack objected. "If you observe any violation whatsoever, it will simply mean that your die is weighted!"

Ted had to agree. Obviously you can no longer continue using science once you have eliminated one of the principles on which it rests. It is like asking: if God is omnipotent, can he make a stone so heavy that he won't be able to lift it? Or can God win an ace with a two? If he can, he is not playing poker. A game without rules isn't much of a game.

In any case, Ted should have known better than to expect support for unconventional ideas from Jack Steinberger. Physicists generally do not veer off the beaten track and Jack even less so than others. But open-mined Pascal also had difficulty at times with Ted's ideas. Pascal was younger than Jack and more approachable. (It could be the Nobel-Prize effect; Ted remembers Jack saying that being awarded the Nobel Prize changes people—he had said this, of course, before getting his prize.)

Ted and Jack were now at CERN, the large European center for nuclear research in Geneva, Switzerland. In the mid 1970s CERN had become the most important laboratory for particle physics in the world. Jack had taken up a post here as division director whereas Ted, having completed his studies at Columbia, had come as researcher and joined a group of physicists lead by Pascal.

The three of them had a somewhat bumpy encounter one morning. Ted and Pascal had been preparing transparencies for a presentation using a thermal transparency copier in which one slides a clear plastic film on top of a Xerox photocopy in order to transfer an image from the paper to the plastic film. Looking carefully at the results they had been startled to notice that one transparency had come out better than its original. The plastic film came out with machine-typed numbers where the original Xerox copy had handwritten numbers; a small but unbelievable improvement! As they wondered in disbelief, Jack walked by so they presented him with this mystery.

"How can a copying machine introduce information that did not exist in the original?" remarked Jack.

"That is exactly what we do not understand," exclaimed the two physicists.

"It cannot be," said Jack as he walked away.

"But, Jack, take a look."

He was not interested in taking a closer look at their evidence. Miracles were simply not an option and therefore not interesting.

Pascal and Ted kept looking at their puzzle for a long time. Finally Pascal noticed that the machine-type numbers could be peeled off from the transparency's reverse side. Apparently, in an effort to beautify the presentation, a secretary had transferred machine-typed numbers from a lettraset sheet and had covered the hand-written numbers on the original. But the heat of the copying process had loosened the machine-typed numbers and transferred them to the transparency film, thus revealing the hand-written numbers on the original. Jack had been proved right—not surprisingly—but he had treated Ted and Pascal as not very intelligent for having become excited by such an issue. And yet, like most physicists, they were intelligent.

Ted always got a kick out of people's reaction when they learned he was a physicist: "You must be very intelligent." He invariably responded by explaining to them that there is always the same percentage of idiots in *any* group of people, no matter how intelligent its members. It is a consequence of the tendency to call an idiot anyone who belongs to the bottom few percent of the group's IQ distribution. Ted did not hold intelligence or science in veneration. On the contrary, he would not hesitate criticizing scientific work even when the researcher's imposing scientific stature forced most physicists to shut up.

Today, for example, he attended the weekly seminar in the main auditorium of CERN. These seminars constitute a first "publication" of new scientific results. CERN physicists attend them regularly in the same way Catholics attend church on Sunday. This time, as happened more often than not, Ted found things to criticize.

"This is not science," he whispered to the physicist sitting next to him. "This guy claims he made a discovery but his conclusion is only *probably* correct. Three standard deviations do not warrant certainty. Discoveries are not like that. Imagine Columbus coming back from his voyage to the New World and reporting to Queen Isabella that it was *very probable* that his boat had survived the roundtrip to the other side of the ocean and was now in the Lisbon harbor loaded with gold and other exotic items!"

The auditorium was packed with physicists but also with journalists and television cameras. The speaker was world-renowned physicist Carlo Rubbia, not yet a Nobel-Prize laureate. He had been expected to make a crucial announcement concerning new experimental results proving the existence of neutral currents, a phantom phenomenon coveted by theoretical physicist for years. The anticipation had been built up not because of exceptional experimental work, but because of the speaker. Most physicists at CERN did not think very highly of Rubbia's experiment, but they all agreed that his word counted in physics. His name carried a lot of weight. It was his reputation that had built up expectations about this seminar.

Rubbia was physically big, brilliant, arrogant, and spoke very fast. His notoriety extended outside physics circles. Once, Ted and his wife were having lunch at the CERN cafeteria when a noisy group of physicists entered.

"Here comes Rubbia," Ted remarked.

"Which one is he?" she asked excited.

"He's the cockiest one in the lot," snapped Ted sarcastically.

But today it was also the presence of the media people that irritated Ted. The CERN auditorium, where physicists gather more than once a week to report on research, listen to criticism, and argue, is a large amphitheater with wood paneling and seats equipped with microphones and earphones. Most of the time it is dark, illuminated only by the light cast on the enormous screen by the speaker's overhead projector. To physicists this place is as a church sanctuary is to priests. But today the place was crawling with journalists and their clutter of cameras, portable light fixtures, and recording equipment.

It reminded Ted of a similar event a few years ago when he was still a graduate student at Columbia University. He had been working on his Ph.D.

thesis when the phone rang and his friend Mihali Yannopoulos told him excited that two physicists would be reporting on Uri Geller that afternoon at Pupin Hall. Ted had been out of touch with activities on campus, but Mihali, an uneasy and inquisitive mind, would not miss an event like this.

• • •

Columbia's physics department boasts a rich scientific heritage. It is housed in Pupin Hall, a 10-story red brick building capped by a green observatory dome. The walls and corridors of the edifice are darkened by the passage of time and its elevators feature sliding iron gates. The atom was first split in an early version of a cyclotron in its basement, giving birth to the Manhattan Project that led to the development of the atomic bomb. The building has been declared a national historic landmark. Its main amphitheater is an older version of the CERN auditorium and serves the same purpose.

When Ted arrived at the amphitheater that afternoon he found it packed with students, professors, journalists, and cameramen. Three Nobel-Prize laureate professors were seated in the front row. Two young physicists were to take turns presenting their findings.

"We have returned to where we graduated only a short time ago," began one of them at the podium, "because we have made some observations that are puzzling and we would like you, our teachers, to help us understand them."

The two physicists had been working at the Stanford Research Institute for some time, and among other research projects they had studied Uri Geller, the controversial magician who had recently made his appearance and was rapidly gaining notoriety around the country. The two speakers proceeded to present a series of experiments they had conducted with Uri Geller and their conclusions. The experiments concerned the bending of spoons, the guessing of hidden drawings, the moving of things at a distance, and the like. Their conclusion was, "We are unable to explain these phenomena."

The heavyweight scientists seated on the front row seemed annoyed, almost angry. They voiced serious criticism.

"Did you test Geller's ability to perceive infrared and ultraviolet light?"

"Did you insert a lead sheet between Geller and the envelopes with the drawings he was guessing?"

"Did you monitor the levels and the variations of all electromagnetic radiation in the experimental environment?"

"Did you measure and classify Geller's muscle response to stimuli?"

A barrage of questions, to most of which the answer was no, they had not.

"Then why are you coming here wasting our time?"

Their anger may have been accentuated by the presence of the media. The publicity about this presentation on Uri Geller surpassed the publicity of all previous presentations in this amphitheater, and there had been many of them and rather important ones. The presence of the television networks in Pupin Hall's main amphitheater for non-scientific reasons, such as the feats of a magician, had been tantamount to sacrilege.

But there was another explanation for the professors' reaction. They seemed to be saying, "You have not done your homework and that is why you have results that you do not understand. Had you been good scientists, you would have proven that there is no magic here." The reverberation throughout the amphitheater was THERE CANNOT BE MAGIC.

• • •

Reminiscing about his student days in New York made Ted feel guilty about a promise he had not kept, a promise he had made to his friends Aris and Mihali before leaving New York for Europe. He had been in Geneva for four years now and had made no serious attempt to locate the secretive Gurdjieff Institute. He did not realize at that moment that his guilt feeling was misplaced, that in fact he had his own reasons to do exactly what his friends had asked him to do, and that his reasons were related to the inadequacies he found in physics.

3 – Precocious Cross-Disciplinarity

Ted, Aris, and Mihali had become friends during graduate school at Columbia University. They were all Greeks and had lived together in New York City through the Beatles' cultural wave and the 1968 student uprising. Over the years Aris had become a shrewd architect with a strong interest in metaphysics. Mihali had become a Leonardo da Vinci with a strong interest in just about everything, and Ted had stubbornly stuck to physics.

They enjoyed each other's company and often spent hours discussing various subjects. Ted brought ideas from physics, Aris from mysticism, and Mihali mostly from the worlds of philosophy and art.

One night Aris suggested that there should be seven steps involved in optimally reacting to *any* event. He was prompted by an incident that day when his reaction had not been the best. "There are times when everything I do comes out right," he said "but at other times my behavior leaves things to be desired. What are the steps to follow to ensure optimal performance *at all times?*"

The three took turns at the subject. At the end they reached consensus but only for the first two steps.

They agreed that the first thing to do, no matter what the situation, is to do … nothing. This eliminates panic reactions and associated erratic behavior. It is the equivalent of resetting everything to zero, so that one can start from the beginning. It permits reassessing the situation by accepting inputs afresh.

The second step is to identify any possible danger within the next three seconds. This was thought to ensure survival from threats toward which one can take action.

But they could not go further. It was impossible to identify any of the remaining five steps, if indeed there were exactly five more. The discussion was left unfinished that day and did not advance further in any of their subsequent get-togethers.

Another night Mihali argued that the climax of the ancient Greek civilization, with its byproducts such as the development of the arts, philosophy, and birth of democracy, was not an isolated event. At exactly the same time on the other side of the earth Buddha was planting the seeds of one of the world's major religions. Both cultures explored the nature of the world and our place in it simultaneously. Both produced refined ideas and profound understandings. Mihali went on to suggest that there must have been secret contacts and cross-fertilization between the two civilizations.

"Secrecy," Aris added, "is always on top of the agenda when it comes to *real* knowledge." He emphasized the word *real*. "Real knowledge," he continued, "is different from academic knowledge in that it can achieve personal gains, yield power, and change the face of the earth. Real knowledge is kept secret and becomes accessible to only a few, as if it were some kind of limited resource, which can be effective only when concentrated on few hands. Recipes, such as that for making Coca Cola, are well-kept secrets. In contrast, academic knowledge is published and becomes available to everyone. Complications arise when academicians, unaccustomed to the tradition of secrecy, get their hands on real knowledge, for example the harvesting of nuclear energy, which then becomes publicly accessible.

The ancient Greek Mysteries—secret cults representing the spiritual attempts of men and women of all ages to deal with their mortality—preserved teachings that contained much real knowledge. Contrary to popular belief, this knowledge did not originate either in Egypt or in Greece, but could be traced to pre-Vedic India. The Greek Mysteries were the last surviving relics of the archaic wisdom enacted under the guidance of the Great Initiates. Herodotus informs us that the Mysteries were introduced into Greece by Orpheus, allegedly the son of Apollo."

Hearing the name of Orpheus, Ted was reminded of the book by Schuré that had so impressed him as a child; Orpheus was one of the great initiates discussed in the book.

"The loss of the Greek Mysteries marked the beginning of the Dark Ages of Europe," Mihali took the relay. "The ancient Greek philosopher Epictetus praised the moral value of the Greek Mysteries, and Plato asserted that their real object was to restore the soul to its primordial purity, that state of perfection from which it had fallen," he said, and paused.

Aris continued from where Mihali left.

"When Christ disappeared between the ages of fifteen and thirty, he most probably joined the Mysteries network. He studied the wisdom therein for fifteen years. But then he went on to do what no person initiated into the Mysteries ever dared do; he went public. The vow of secrecy was sacred and the punishment for disobedience was of the utmost severity. In his comedies, Aristophanes makes sure to precede the uttering of any indiscretion with the phrase: 'Because I have not been initiated, I can say whatever I want.' But Christ had been initiated and so he was punished by crucifixion.

"Drawing on the secret knowledge of the Mysteries," Aris continued, "Christ was able to achieve a phenomenal success. For two thousand years, millions of Christians all over the world have been beaming good wishes to him. Sending good wishes does more than keeping one's memory alive. It nourishes the astral body."

The existence of the astral body as a continuation after death and a replacement of our physical body was one of Aris's pet theories.

"By the way," he added, "during the last supper Christ couldn't possibly have offered bread and wine to his disciples as the Bible says. The stakes were much too high for symbolic gestures at that moment. He offered them the real thing, his own blood and part of his own flesh. It has been documented in magic literature that blood connection facilitates contact of the astral bodies, something that Christ would have later need for. Some of this thinking may also lie behind the custom of becoming brothers in blood by cutting and touching bleeding veins. The church is not entirely out of this loop. In the mountainous villages of northern Greece the priest supervised the execution of this custom in older times. The symbolic wine-and-bread ritual for communion was introduced later by the church to facilitate things for early Christians, just as was the abolishment of circumcision. This way the ordinary layperson could easily emulate the ritual Christ had gone through with his disciples."

And so continued the intellectual introspective get-togethers of Ted, Mihali, and Aris, sometimes producing titillating pearls of wisdom. In one meeting Mihali dominated the conversation by quoting Socrates in his famous "all I know is that I know nothing." But he gave the quotation in ancient Greek, "Ἕν οἶδα, ὅτι οὐδὲν οἶδα," and pointed out that an accurate translation could be "one thing I know, nothing," which is also amenable to a different and more far-reaching interpretation. Socrates may have intended to encrypt in this saying that there was one thing he understood well and that was nothingness or the absolute zero.

Mihali was well aware that he could be over-interpreting Socrates's remark, but used this idea as a platform to launch a philosophical discourse linking zero to the complete absence of anything and oriental approaches to meditation that lead to nirvana.

Ted found it fascinating that Socrates may have been interested in the complex notion that physicists refer to as the ideal vacuum. He picked up the thread and explained that the complete absence of anything is a theoretical concoction, which in reality does not exist. One pragmatic reason is that any vacuum chamber has walls made out of some material that will emit photons in the form of black-body radiation. The energy of these photons depends on the temperature. When this soup of photons is in thermodynamic equilibrium with the chamber walls, the vacuum can be said to have a particular temperature, as well as a pressure. One may then argue that the complete absence of anything can be achieved only at the temperature of absolute zero.

But even then there would be other reasons for which absolute vacuum cannot exist. One reason is linked to Heisenberg's uncertainty principle, which states that no particle can ever have an exact position. A particle in physics exists only as a probability function of space, which has a certain non-zero value everywhere in a given volume. The space inside an atom, or the space between molecules, is not a perfect vacuum because there is always *some* probability that there are particles there.

Finally and more fundamentally, quantum mechanics predicts that vacuum energy can never be exactly zero. Even at the temperature of absolute zero, the lowest possible energy state will be the zero-point energy that consists of a seething mass of "virtual" particles that have brief existence. This is called vacuum fluctuation. In fact, some of these "virtual" particles can be charged particle-antiparticle pairs, such as electron-positron pairs, which could be reoriented in the presence of an electromagnetic field. This reorientation is referred to as vacuum polarization.

More recently it has been hypothesized that large amounts of dark energy —an unknown and unobservable form of energy—accounting for 68% of the total energy in the present-day observable universe permeates all of empty space.

In a nutshell, the complete absence of anything has lots of energy, pressure, depicts fluctuations with the creation of particle-antiparticle pairs, and can be polarized. If Socrates had even a suspicion of any of this, he would indeed be the wisest of men.

In their next get-together it was Ted's turn to lead the discussion. He had read an article in *Time* magazine presenting high-IQ sperm banks as genius factories capable of raising the population's average intelligence. Ethical and philosophical issues were discussed in the article, but the possibility that society may at some future time consist only of geniuses made him worry for a reason not mentioned in the article. There is a law of physics that forewarns of a hidden danger if all people are given the chance to have *maximum* intelligence.

"Nature favors the release of captive energy," Ted began. "Elevated items tend to fall down to the lowest possible level. Heated items will cool down and refrigerated items will warm up. Following the release of energy, the final state enjoys more stability and its elements are more tightly bound.

Downstream of a waterfall the water is calm and more difficult to displace. The lower the energy of a final state the more stable the state will be, and the more tightly bound its elements. One may say that in nature things tend to arrange themselves through the action of some force (gravity in the case of the waterfall) so that the minimum-energy configuration is attained.

"Extracting and using the energy trapped in a certain configuration destroys potential differences and produces a flatter more uniform 'land-scape.' However, the new state, rid of energy, becomes more tightly bound and stable. Contrary to intuition, two things tightly bound to each other, have low energy content. Just think of the fact that in order to separate them you need to *add* energy to their system; the more so, the more tightly bound they are. Thus, for things that can 'fit' well together, there will be forces pushing them together stemming from the simple fact that their united state will have lower energy. However, this natural matching tendency may favor embar-rassing social situations.

"Imagine that sometime in the future, through technological advances in genetics, people can pre-program their children to be at the top of the scale concerning intelligence, looks, or any other variable. The distribution of the population in this variable will no longer look like the normal bell-shaped curve depicting the familiar percentages of poor, average, and exceptional individuals. It will be a very narrow distribution like a spike at the top of the scale, with practically everyone being equally brilliant, other than a few mis-takes and accidents, a small percentage of individuals that for technical reasons did not make it to the top. With everyone a genius, there will be no exceptional people; no one would stand out. But there are some roles in society that are reserved to a small fraction of the population *by definition*. One of them is leadership; it belongs to precious few exceptional individuals. The only exceptional people available in this fictitious society would be the few "mishaps" stemming from errors and accidents in the programming process. The mere fact that leaders need to be *few* and *different* from the masses may push leadership roles in the hands of the blunders!"

Ted wanted to write a letter to the editors of *Time* magazine suggesting these ideas. Mihali advised against it. "They are not the type of intellectuals who can appreciate thinking along these lines," he said and proved right. When Ted wrote his letter anyway, the answer was that the magazine's science editor "could not see Ted's point." Ted decided to waste no more time with the mass press and concentrate instead on the weekly meetings with his two friends.

4 – The Order of Mendios

One day Mihali told the other two that there was a book they had to read: *In Search of the Miraculous* by Peter Ouspensky, a 20th-century Russian philosopher and mathematician. The book fit well with Aris's and Ted's predilections. The former was excited about the miraculous aspect. The latter felt reassured that the investigator was a bona fide scientist (a mathematician). Mihali had unearthed this book as he tracked down footnote references in a bestseller of the time: *The Master Game* by biochemist and drug expert Robert de Ropp.

Ted read the huge volume from cover to cover, studying it meticulously as if it were physics textbook. Ouspensky recounts that while he was researching the question of the possible existence of miracles, he came across a dubious character named George Ivanovitch Gurdjieff, who changed his life. Gurdjieff had purportedly tapped into secret knowledge that he had discovered in central Asia, and in particular in a remote monastery high in the Tibetan Himalayas. He supplemented it with western thought and came up with an elaborate approach for "the harmonious development of human beings." The approach transcended the limits of science, religion, art, and psychology. Despite the veil of secrecy under which Gurdjieff operated, the France-based institute drew interested people from all around the world, and by the 1930s boasted a sizeable following that included among others renowned architect Frank Lloyd Wright and English writer Katherine Mansfield.

Much material in that book spoke directly to Ted's heart. In it, science was not negated but augmented. Here is an example: the approximately 100 chemical elements tabulated in Mendeleyev's celebrated periodic table suffice to describe the composition of all matter on earth. But Gurdjieff argued that this classification is inadequate. There is a more complete way to organize matter. Take water, for instance. Its chemical name is H_2O, indicating that water is made of hydrogen and oxygen. But water can exist in three different states. In its liquid state water is indispensable for sustaining life. Not so in the ice state, which may preserve life but is not very friendly to it. Finally, in the state of steam, H_2O is utterly hostile to life; in fact it kills it. From this point of view, using a single symbol for water irrespective of its state appears a sloppy oversimplification.

Gurdjieff's taxonomy has many more elements than the basic 100. The more "refined" a substance is the smaller the number of its sequential position.

Solid food resides at level #768 whereas water occupies position #384. The air we breathe is in position #192. Hydrogen, which occupies place #1 in Mendeleyev's table, occupies #12 in Gurdjieff's table. This kind of classification of matter takes into account its relationship to life and the functions of the human organism.

Ted appreciated this classification even more when he realized that periodicities now became exact. In Mendeleyev's periodic table the halogens (Fluorine, Chlorine, Bromine, and Iodine) occupy positions #9, #17, #35, and #53 respectively, defined by their atomic numbers. The halogens have similar properties because their atomic numbers are *almost* in ratios of 2. In Gurdjieff's classification these elements occupy positions #24, #48, #96, and #192 respectively, which display *exactly* factors of 2." The slight inexactitude [in Mendeleyev's classification]," writes Ouspensky, "is brought about by the fact that ordinary chemistry does not take into consideration *all* the properties of a substance, namely, it neglects the 'cosmic properties.'"

The idea of expanding the realm of hard science to allow for the inclusion of such notions as consciousness, feelings, the purpose of life, and God awoke in Ted old questions and hopes for understanding issues, which he had abandoned while sticking to beaten-track physics.

Ouspensky's book provided the three friends with a new host of topics for their weekly get-togethers. At the same time, their meetings became more serious, took place during daytime without alcohol or drugs, and took on some ritual. Now they would sit down on the floor at the apexes of an equilateral triangle drawn on a round rug. They would take turns in speaking and the speaker could not be interrupted unless both of the others agreed to the interruption. It was Aris who designed and imposed such rules. In fact, one day he declared:

"I have been empowered to initiate you to the Order of Mendios. Are you guys willing to join?"

The others were taken by surprise. Who is Mendios? Who gave Aris the power to make initiations? What would it entail to join this order?

Aris brushed aside their questions. "You will have to take risks for whatever you do not know. What I am asking is whether you are willing to join or not?"

Mihali and Ted were not about to take him very seriously but they agreed in order not to disappoint him.

"In that case, next week each one should bring some dirt from in front of his house and a small jar with virgin olive oil," he instructed. "We will then go through the initiation ceremony here."

The ceremony took place in Aris's living room, which had been emptied of all furniture. It lasted more than one hour. Two typed pages had been prepared for each one of them. One included the steps to be followed and the other was text—much of it in verses—that should be read in a loud voice, one section at a time at Aris's prompt. It was a hymn to Hermes, the ancient Greek god of commerce, communications, and magic. At one point the ritual called for all three to strip naked and walk around in the empty room reciting the verses. Aris took it very seriously so the other two complied obediently. After all, a feature they had recently added to their group's activities was the speaker's right to impose exercises on the other two. At the end of the initiation they held discussions.

For several more weeks the meetings of Mendios centered around topics from Ouspensky's book, until one day Aris put on a grave air and announced: "I have located a chapter of Gurdjieff's Institute in New York City."

"Is there such a thing? Where is it? How did you find it?" Came a barrage of questions.

"I put an ad in the *Village Voice*, explained Aris.

"What did you put in this ad?" asked Ted, who knew of Gurdjieff's preoccupation with secrecy.

"I said that I wanted to establish contact with G.'s Institute and offered my telephone number. Someone called me a few days later. We met in a bar and he invited me to one of their meetings."

Ted and Mihali wanted to know more, "So what happened?"

Aris paused and then said emphatically, "What we are doing here is far more advanced than what they are doing there."

The intensity diffused. They concluded that they were not missing out on anything significant and decided not to pursue Aris's discovery further, but to continue carrying on the way they had done up to now.

Months went by and much ground was covered during their meetings in the Order of Mendios. But there came a moment for the trio to split. Ted had accepted a job in Geneva to carry out experiments in particle physics at CERN, Mihali was going to try his luck in Saudi Arabia, and Aris would remain in New York City to work as an architect. Nevertheless, they pledged to maintain allegiance to the Order of Mendios.

As they bid each other goodbye, Aris made Ted promise that he would try to find Gurdjieff's Institute in Europe. He felt that Geneva was a likely place for such activities because of its proximity to Fontainebleau, where Gurdjieff had lived and worked. He also expected that they would be doing more authentic work in Europe than in the New York chapter he had visited. Ted had agreed to seek them out.

5 – Hypnosis Demystified

It is not wholly accurate that during his first four years in Europe Ted made no attempt to locate Gurdjieff's institute. He made two attempts on different occasions. The first try was during his early days in Switzerland, when he loved to drive around exploring the mountainous countryside with its beautiful lakes and breathtaking views from high in the Alps. He had read in a catalog of esoteric literature (*A Spiritual Directory for the New Age*) that in Chandolin, an alpine village in the canton of Valais, Gurdjieff's followers had an establishment where they held summer camps. It was a good excuse for Ted to drive there one day during his first summer in Switzerland.

At 2000 meters Chandolin is the highest Swiss village that is inhabited year round. Off the great valley of the river Rhone, Chandolin is perched on the side of a mountain across a smaller valley from a skyscraping alpine range. Ted stopped at a café in the village and with his broken French asked the waitress whether she knew of a Gurdjieff institute in the region.

"Come again?" replied the waitress showing confusion. Knowing the secrecy Gurdjieff followers cultivated, Ted hesitated but repeated the question.

"I know no one with this name," she replied.

And yet the village was very small and Ted was sure that they were there.

Ted's second attempt to locate them was a year later when he decided to do as Aris had done in New York, and put an ad in a paper. He chose *The International Herald Tribune,* which was widely read in Geneva, hoping for a response from an English speaker. Following Aris's example, he composed a short and cryptic announcement:

> Searching to establish contact with a
> G. Institute in the greater Geneva area.
> Please call 022-968809 (evenings).

The following week he received a phone call. It was editor Robert Feldman from the offices of *The International Herald Tribune* in Lausanne. Ted answered the phone somewhat apprehensive.

"We have received the announcement you filed with us last week but I regret to inform you that we do not carry sexual ads in our paper."

"What do you mean sexual ads?" Ted protested.

"Well, what does G. Institute stand for?"

"The G. stands for Gurdjieff, a 20th-century Russian philosopher," Ted rushed to explain.

"That's strange," said Feldman, "I took philosophy as a minor in college but I never heard of this philosopher. Anyway, we will pass the ad."

Ted never received the phone call he was hoping for, and he made no other formal searches. But all his friends came to know of his fascination with Gurdjieff. He would often quote Gurdjieff in social gatherings to impress interlocutors with unusual and clever ideas. A typical exchange would go as follows:

"Oh, it's all ready midnight! I cannot believe we've been here four hours. It feels as if we came only one hour ago," says Daniella preparing to leave Ted's dinner party.

"It simply means that you have spent only one hour of your own clock time, while the clock on the wall indicated four hours," interjects Ted, quoting Gurdjieff's internal-clock model.

"Who is this Gurdjieff and what does his internal-clock model say?"

Ted takes the opportunity to make a show.

"We are all given at birth the same quota for a lifetime, let us say 80 years. However, our clocks do not all run at the same rate. Some run faster and others slower than the mechanical clock on the wall. Moreover, the rate of our internal clock may change from one situation to another. If someone has stood you up on a date, time moves slowly. You keep looking at your watch and five minutes may seem like half an hour. In this case, you have aged by half an hour while the watch on your wrist advanced only five minutes. And the contrary, when you are having such a nice time that time seems to fly by, you are aging less than the clock on the wall indicates. At the end, we all die 80 years later, but some of us consume these years in 70 or 60 calendar years, while others die when the clock on the wall has counted 100 years. The way to tell how fast your internal clocking is ticking is to be sensitive to how time feels to you as it goes by."

The next party was at Daniella's place and Ted was excited to find out that her husband, Fernando, purportedly knew how to hypnotize people. Ted had never forgiven himself for wasting his opportunity to be hypnotized by a professional during his days at Columbia.

• • •

Ted was still in graduate school when his classmate Steve told him about a hypnosis course to be given for the first time to medical students at Columbia's Presbyterian Hospital. The course aimed to train future doctors on using hypnosis as a general tool for various medical treatments ranging from losing weight and stopping smoking to non-invasive anesthesia during labor and even surgery. Ted and Steve decided to attend the course as auditors.

During the first session the professor had asked for volunteers to demonstrate his technique. Ted had hesitated, but when he later saw the spectacular power of hypnosis on those who did volunteer, he regretted missing the opportunity. He wished he had experienced hypnosis first hand, because some of the behaviors he witnessed on stage during the several-week course were indeed incredible.

One day the professor told the class that his guest that day was a very hypnotizable subject. The professor had brought with him a man wearing thick-lens glasses and when he asked him to remove them the man was unable to read the clock at the far end of the hall. But a little later under hypnosis—which required only a single jest from the professor—he was able to read the clock with no difficulty. It was Ted's first witnessing of an unquestionable manifestation of mind-over-body dominance.

• • •

Now at Daniella's party Fernando was going to become Ted's second chance. Fernando suggested that they isolate themselves in the bedroom far from the others and the noise. Ted was all too willing. Fernando asked Ted to lie down, relax, and take deep breaths. Ted was familiar with the procedure. Fernando kept talking to him in a soft voice trying to make him relax further. Later he asked Ted to close his eyes and proceeded to suggest that his eyelids would get so heavy that he would not be able to open them. Some time later Fernando declared that Ted was no longer able to open his eyes. Instantly Ted opened them. Fernando was vexed.

"Why are you opening your eyes? Are you trying to test me? Are you here for an experiment or for being hypnotized?"

Ted had to agree that he wanted to verify whether indeed he was unable to open his eyes. This verification seemed essential to Ted. Not to Fernando. He accused Ted of not cooperating and gave up on the hypnosis session.

They walked back to the living room where the others wanted to know of the results.

Fernando blamed Ted for the session's failure and attributed it to Ted's playing the scientist and having a secret agenda other than honestly wanting to be hypnotized. Ted admitted wanting to verify the effects of hypnosis and then he recounted a scene in the hypnosis course he had taken at Columbia.

"One day," he said, "the professor brought in class a man that he referred to as the devil's advocate. He was another professor whose thesis was that there is no such thing as hypnosis. The new professor took the podium and prepared to make a demonstration of his own. He asked one student in the front row to give him her right shoe, another one to give him his glasses, and a third one to give him his wallet. The three students complied. The professor then turned to the class and proclaimed, 'This is what hypnosis is all about. Did you ever think it would be so easy to get hold of someone's wallet, glasses, or shoe?' The moral of the story is that there is a thin line between full cooperation and hypnotism, a line that is perhaps fuzzy and ill defined. There maybe no essential difference between *really* wanting to obey a command and being unable to disobey it. And then, of course, there are people who are more susceptible to suggestion, that is more hypnotizable, than others," concluded Ted.

He did not talk about it then, but he made a mental link between hypnosis and Gurdjieff's fundamental premise that the greatest human weakness is being highly influenceable. Television and other media regularly shape everyone's notion on most issues ranging from what constitutes beauty to ideas that may lead to terrorism. He was comforted by the thought that scientific facts are to a large extent "immune" to what individuals may think.

Despite the unsuccessful hypnosis experiment at Daniella's party, Ted felt relieved from his regrets for having missed a chance back in the hypnosis course. The phenomenon of hypnosis was largely clarified for him now and he could reclassify it in his "weirdoes cabinet," his mental filing cabinet whose drawers were labeled with question marks. Some drawers were labeled with many question marks; others with few. He filed in these drawers hard-to-believe claims and things outside the realm of science. Things such as Uri Geller's feats, Steven von Daniken's writings, UFO sightings, Lobsang Rampa's opening of the Third Eye, miracles, supernatural events, mysteries, and in general everything science cannot explain and he could not outright discard. The cabinet served him as temporary storage that he visited and rearranged from time to time. Ted's weirdoes cabinet also contained some of his own first-hand experiences and experimentations.

6 – A Weirdoes Filing Cabinet

While at graduate school, Ted and Steve had shared more than being class-mates and doing their thesis research on the same experiment during their years at Columbia. They both featured long beards and drove Volkswagen beetles, which they serviced and repaired themselves. Above all, however, they were both hard-core scientists with a penchant for the unexplained and a flare for provocation. When they were invited to a birthday party of a somewhat uptight individual, they showed up with a dozen inflated party balloons. The party's agitated crowd did not notice that only half of the balloons pointed upward as balloons filled with helium normally do. Ted and Steve had filled the other six with nitrous oxide (N_2O), otherwise known as laughing gas, that they had obtained from a gas cylinder at the Brookhaven stockroom. At a moment that seemed appropriate Ted suggested that they try to speak like Donald Duck by inhaling the helium content from the balloons. Men and women raced for the balloons while Ted and Steve laid back and enjoyed watching their comical behavior.

Although this was an era of flower children and drugs, people at the time were not generally aware of the properties of laughing gas. Steve had discovered an obscure collection of essays on the practice of inhaling nitrous oxide, one of which had been written in 1882 by psychologist and philosopher William James, the brother of writer Henry James. Nitrous oxide provokes a short and intense high, stimulating convoluted thought processes. At small doses it makes people laugh and act silly. The party in New York became legendary. Years later Ted astounded a drug user, who had already been in Geneva's jail for drug possession, by revealing to him that nitrous oxide could be procured legally. Geneva's grocery stores sold it in pressurized cartridges destined to make whipped cream!

One day Steve found out about an unusual conference organized in a large hotel in lower Manhattan. He and Ted rushed downtown to listen to a number of para-scientists make audacious claims. One of them, Robert Monroe, gave detailed accounts about his wanderings outside his body, as documented in his freshly published book *Journeys out of the Body*. A couple of other "scientists" had studied eggs and reported on electric signals picked up from eggs whenever other eggs were being broken in a nearby kitchen for omelet purposes. And a large session was devoted to the existence of the aura or "life force" that allegedly surrounds each living thing and that some people can see. In fact, Kirlian photography is capable of capturing this aura on film. More than that, someone claimed that a Kirlian photograph of a tree leaf torn in half can yield an image on film of the whole leaf. This could presumably be explained via paranormal forces.

Steve and Ted listened distrustfully. They felt as if the people around them belonged to a different species. The two physics students realized that in their lab they had access to all the equipment necessary to detect auras and to take Kirlian photographs. So they decided to experimentally challenge some of those claims.

The recipe for Kirlian photography involved a photographic plate and a high-frequency high-voltage electric field. That weekend Ted and Steve went to Columbia's NEVIS laboratory—where they prepared the experiment for their theses—and set up the high-voltage generator with Polaroid film plates. Ted volunteered his thumb because he knew that high voltage is not dangerous at high frequencies.

They experimented with a dozen different conditions of voltage and geometry. There was often an image left on the film outlining Ted's thumb, but the explanation was anything but magical. Small insensible discharges between the finger and the photographic plate would leave an imprint of the finger outline. But there was no way that this procedure would leave an imprint of a whole leaf from a leaf torn in half.

As for "seeing" the aura, they hooked up the lab's most powerful photo-multiplier, a huge sensitive electronic "eye" capable of "seeing" single photons. In a dark room it revealed no signal around Steve's or Ted's head, no matter how hard the two men tried to detect one. The topics of aura and Kirlian photography were dropped out altogether from Ted's weirdoes cabinet of phenomena "waiting" for explanation.

But Monroe's out-of-body experience survived. Not simply because it proved too difficult to experimentally dismiss it but also because Ted had something of a partial confirmation. One summer evening something weird happened when he went to bed. He had turned off his radio before going to bed as usual but a little later, as he relaxed and waited to fall asleep, he heard the radio playing. Being thoroughly convinced that he had turned it off, he was shocked by the fact that it was now playing. It constituted an impossibility. The conflict became irresolvable and began terrorizing him as if he was witnessing something out of this world. The terror built up to the point of making his whole body shake resulting in an uncontrollable trembling. It was at that moment that he realized that he could sit up in bed while his body was still lying down immobile. He remembers sitting up and turning around his head to see the upper part of his body still lying down in bed. His terror subsided and he began considering getting out of his body altogether. But he couldn't. The only move he could make was go back to his previous lying-down position. He did that and things returned to normal with the radio not playing any more and his body feeling as though it had not moved at all for a long time.

This was Ted's only paranormal experience along the lines of Monroe's stories and he was not able to reproduce it despite multiple attempts. He even bought Monroe's book and followed the directions therein to no avail. Still, he kept the topic in his mental cabinet in a drawer with a reduced number of question marks.

He described his experience to Aris and Mihali in one of their Mendios sessions. They listened to him with compassion. They wanted to know whether Ted had been under the influence of alcohol or any drugs. He told them he had not. They could not give explanations but did not dismiss the phenomenon either. As a matter of fact they would have accepted much more. They were not hard-core scientists.

At about the same time another item became reclassified in Ted's weirdoes cabinet: the paranormal near-magical experiences of Carlos Castaneda recounted in his book *The Teaching of Don Juan,* which popularized American-Indian practices with hallucinogenic plants. Ted had read the book mainly because Castaneda, like Ouspensky, was a rational and scientifically minded researcher. But Ted had classified many of the events described in it as weirdoes. Then came news about Castaneda's second book, not yet available at the bookstands.

Mihali summoned Ted one evening to a surprise visit and lecture by Carlos Castaneda, who was going to talk to Columbia students about his new book. Poorly advertised, the lecture was attended by less than twenty students in a small classroom. Mihali and Ted were there.

Castaneda was anything but an impressive figure, short, rather round, calm, smiling, and benevolent looking. He talked about his new book and the power of seeing. He explained how he got in trouble under Don Juan's guidance when he began seeing people appearing as luminous eggs with tentacles. He ended up visiting an eye-doctor about it. The doctor told him there was nothing wrong with his sight and referred him to a psychiatrist.

Castaneda, always smiling, recounted his visit to the psychiatrist. "By now," he said, "I knew I had to be careful. So I did not tell him that I saw people as luminous eggs. Instead I told him what he wanted to hear. I know what behavior is considered normal so I put it on for him. But deep inside me I knew better!"

Having met Castaneda in person Ted went back to review the fillings in his weirdoes cabinet. Castaneda proved to be not only scientific but also intelligent and cunning. Most importantly he seemed solid and calm, as someone who has found his way, and knows how to get his questions answered. This rendered value to his writings, whether they were true or fictitious.

There were many other items in Ted's mental filing cabinet. New York is an enormous potpourri of oddities and trades. Everything can be found there, particularly if you have friends like Mihali, Aris, and Steve constantly scouting the terrain. Ted skimmed only the cream at the top. His studies made stringent demands on his time. For the most part he contented himself with listening to the others' adventures as they wandered in and out of schools and teachings such as scientology, dianetics, Rimpoche, Maharishi, Tai Chi, Yoga, Timothy Leary, Baba Ram Das, biorhythms, and other movements, most of them in an early stage of development at that time.

7 – Making Contact with the G. Institute

The hypnosis session at Daniella's party may have yielded unexciting results but there was an unexpected pleasant surprise for Ted later that evening when Daniella told him that she had recently met Jean-Paul Rufenacht, who knew of a Gurdjieff group active in Geneva. She had talked to Jean-Paul about Ted's quest and he had volunteered to meet with Ted and answer all his questions. Ted could finally fulfill the promise he had made to his brethren in the Order of Mendios.

Jean-Paul and Ted met for a drink after work. Indeed there was a chapter of Gurdjieff's Institute in Geneva. It was run under the aegis of Michel de Salzmann, spiritual leader Number 1 in the hierarchy of the Paris-based Gurdjieff Foundation. Jean-Paul had been a member of the Geneva chapter for a couple of years, but had later severed ties with it. His job left him no flexibility to attend all of the group's activities and Michel refused to give his permission for partial attendance. Still, when Jean-Paul saw Michel for the last time and told him of his decision to discontinue his membership, Michel responded with, "One cannot break away from the Work." Jean-Paul had been somewhat distressed. He understood Michel's remark as not being able to forget the knowledge he had acquired. But the whole thing had overtones of the original sin; once you lose your innocence by partaking in the fruit of knowledge, there can be no going back.

All this did not seem to bother Ted, however. He was glad to have finally found names and telephone numbers. He thanked Jean-Paul and lost no time. He called and asked for an appointment with Michel de Salzmann. He got one for the following Saturday in Geneva's chic old Hotel de la Paix.

Ted did not know what Michel looked like, but he had seen many pictures of Gurdjieff. When he entered the hotel lobby he saw a man sitting on the couch at the far end of the hall who strikingly resembled Gurdjieff. People aging in couples tend to look alike; why shouldn't devout followers look like their spiritual leader.

Michel had a number of questions for Ted. What was his background? How much did he know about Gurdjieff? Who told him about the Geneva group? Why did he want to join their work?

Ted was not proud of his answers but Michel seemed to be satisfied.

"I find Gurdjieff's ideas very interesting," said Ted.

"We do not accept tourists," responded Michel, "curiosity *per se* does not qualify as a good motive. You must be feeling disappointed in your life, deceived by your accomplishments, and tormented by questions you cannot answer," and then continued talking about the summer camps they held at Chandolin as if Ted had met the conditions.

With summer approaching, Ted asked whether he could join one of the summer camps.

"There will be in fact a group of beginners," said Michel thoughtfully, "but they are all young; much younger than you. On the other hand, you are not ready for the camp with people of your age. We should wait until next year."

Ted was frustrated. He would have to wait one more year before joining a summer camp in Chandolin. He was less eager to attend weekly meeting during the winter in Geneva. From Ouspensky's writings he knew that it was the intense live-in conditions that were conducive to produce impressive results. The only "miracle" Ouspensky ever described in his book *In Search of the Miraculous* took place in an intense living-in environment. Several days of powerful exchanges and demanding exercises had lead to an incident where Gurdjieff communicated with Ouspensky telepathically but explicitly.

Ted proved right. Not much happened during the following months. He attended meetings regularly, but the members of the local chapter proved unexciting and there were only precious few occasions to interact with Michel, who rarely came from Paris to guide the Geneva group. But Ted had now added a new activity to his Saturday routine.

8 – Running after Entropy on Saturday Morning

Until recently Ted used to devote most of the day of Saturday to what he called "running after entropy," i.e., catching up with overdue errands and preventing things from drifting into disorder. He knew that "things" have a tendency of doing just that. Disorder naturally fills our lives. Dirt and dust will slowly diffuse everywhere in a room. Batteries will lose their strength, light bulbs will burn out, supports will weaken, and loose cables on the floor will get tangled in someone's feet. The environment constitutes a natural niche for disorder and chaos to grow into. It is one of the most fundamental laws of physics, the second law of thermodynamics, which stipulates that the entropy (or disorder) in a closed system will always increase. In simple language this would translate to "things will always get worse."

It was somewhat depressing that such a fundamental law of physics carried so pessimistic a message. Ted's attitude toward entropy became that of preventive and corrective maintenance: stock up on spares, archive documents, repair, rebuild, glue, restore, and so on. Like a pilot in a cockpit he understood the necessity of putting real effort to bring everything back in line so that the set course can be maintained.

But the second law of thermodynamics has a further disturbing aspect hidden in the phrase "closed system." Order can certainly be increased locally, just as Ted and other people tend to do on Saturdays, but only at the expense of increasing disorder elsewhere. When Ted puts his room in order, the rest of the apartment becomes more disorderly. If he put the whole apartment in order, the rest of the building, the yard, and the street in front, where all the garbage and things thrown out piled up, became more disorderly.

Ted wondered whether Switzerland being so orderly (a country that works like a clock) is at the expense of the other countries around it. It seemed to be the case at least with the banking system. Secrecy of bank accounts was progressively abandoned in the European Union but lingered on in Switzerland. This resulted not only in funds hemorrhaging out of E.U. countries and into Swiss banks but also in Switzerland's stubborn refusal to join the E.U. lest it loses this competitive advantage. Instead many complicated bilateral agreements were put in place between Switzerland and each European country, a real mess.

But then most countries around the world do similar things. From the US to China, governments of powerful nations try to increase their internal order (by consuming energy and natural resource) to the detriment of other countries where disorder is increased either by impoverished land and a mess left behind, or by increased pollution.

As the whole world progressively becomes aware of the law that entropy in a closed system will always increase—if not in this formulation—attention shifts toward trying to extend the limits of our closed system to encompass all of the earth. Big-picture visionaries advocate an increase of order on earth at the expense of disorder in outer space. There are proposals to dump nuclear and other wastes in outer space.

Ted's preventive maintenance was aimed at restoring order and thus eliminating potential accidents. Disorder breeds accidents. He felt that all possible accidents are just sitting out there patiently waiting to be materialized. As if by some variant of survival of the fittest the most "agile and competent" accident would seize the first opportunity to slip into the real world. It is a direct corollary of the ergodic theorem.

In mathematical jargon, *ergodic* is an adjective meaning having nonzero probability. Physics, true to its tradition of grafting mathematics onto real-life applications, has created the ergodic theorem, which claims, in simple language, that situations that are theoretically possible will occur if given enough time. By *possible* one can literally consider all imaginable situations that do not contradict fundamental laws. *Enough time* may signify that some situations may take practically forever before they come to life. Nevertheless, the theorem can be proven not only mathematically, but also practically in everyday experience.

Consider fine glassware; its weakness is that it can break. Owners of fine glassware may take particular care, which can substantially extend the lifetime of a cherished object, but one can be sure that sooner or later, something that can break will eventually break. It is only a question of time. This may sound like Murphy's popular law, which states that whatever can go wrong will. But Murphy must have been in a cynical mood when he created it because his law lacks scientific rigor. The true ergodic theorem is amoral, inasmuch as it does not discriminate between good and bad, desirable and undesirable.

Another application of the ergodic theorem is a fact that is common knowledge among business managers. People at their jobs tend to do what they can rather than what they were hired for. Some companies have become successful by allowing their employees more freedom to shape their own jobs. Employees are happier that way. The law says that if they can do something, they will eventually do it, which means there will be a force—a desire—driving them in that direction. Blocking the realization of that desire can result in frustration, which may or may not be openly expressed. We are constantly faced with a multitude of things we can do. Our desire to do them is usually inversely proportional to the difficulty involved.

Obviously, according to the law of large numbers, the more unlikely the event, the longer it will take statistically before it occurs. Imagine an enormous brightly colored moth flying in through the open window at night to land on your arm. Such an event may require waiting longer than your lifetime. In that sense, it may never happen, but had you been able to sit there and wait in front of the window for thousands of years, you would eventually be guaranteed such a visit.

Ultimately, one might expect "miracles." As a physics student Ted had to calculate how long one has to wait in order to see a pencil fly! Since atoms in solids are vibrating in random directions, it is conceivable that there may be a time when all the atoms of one pencil happen to direct themselves upward, in which case the pencil will lift off on its own. Ted's calculation showed that had someone been watching a pencil without interruption from the day the universe was created, he most certainly would not have witnessed such a levitation. In fact, it is much worse; in order to have a fair chance for such a sighting, one would have to wait another period of that length, but this time with every second stretched to the age of the universe! Yet, the possibility is there, and it is only a question of time.

So Ted's running after entropy on Saturdays only postponed rather than eliminated accidents. But now in addition to running after entropy, Ted also had to attend Gurdjieff group on Saturday mornings. He thought that doing so may also be equivalent to some sort of preventive maintenance, but he couldn't quite put his finger on what type of accident he would be thus postponing.

9 – Why Is the Sky Dark at Night?

That year Ted made plans for summer vacation so as to leave a week in late July for the Gurdjieff camp. When the time came, he loaded his car with a sleeping bag and other mountain gear and drove up the Alps for his first camp with Gurdjieff enthusiasts from all over the world. He felt excited and full of anticipation. He smiled as he thought of Mihali and Aris. He would have much to tell them next time they met.

At Chandolin he asked directions for *Le Zoc*, the name of the three-chalet complex where these summer camps took place. It was known in the village as the place where psychiatrist Michel de Salzmann brought his patients for a change of air. And that is why people in the village had never heard of Gurdjief's name. The wooden chalets were situated on a steeply sloping mountainside about fifty meters below the village and could be accessed only on foot. As Ted approached he was engulfed in a breeze of rarefied air, mountain smells, and silence. From a distance he discerned the slowly moving silhouettes of men and women on a large terrace. When he reached it he was told to go to the small chalet. A thin tall man showed him the way.

The small chalet consisted mostly of a tiny room with five crowded beds and a wooden balcony facing an imposing mountain range across the valley. Ted left his sleeping bag on one bed and went back to the terrace where more people had gathered. They were doing nothing, not even talking. Ted approached a friendly looking man with a mustache.

"Hi, this is my first time here. Have you been here before?"

"Oh yes. I have been in the Work for nineteen years," said the man with a Spanish accent. He was Mexican.

"Nineteen years!" exclaimed Ted in awe.

The man rushed to reassure him. "Don't be impressed. Whether you have been doing this for nineteen years or for just one year makes no real difference."

At dinnertime someone rang the bell hanging in the entrance of the largest chalet. They all gathered in the great hall inside it. There was practically no furniture. The floor and the walls were entirely covered with somber oriental carpets. A huge window on the south side presented a magnificent view of the mountains. On the floor square pillows had been arranged in large semicircles facing east. Ted and sixty or so men and women came in and sat down cross-legged, each on a pillow, facing the east wall along which bigger pillows awaited the elders. Everyone waited in silence. In a few minutes the elders walked in a single file led by Michel, and sat on the floor facing the crowd. They included Michel's wife, José, who sat on his left side. After more silence Michel spoke.

Without much introduction or welcoming comments he reminded everyone of the daily program and some of the rules. How they would be woken up at six in the morning, and how throughout the week the Work would go in parallel with ordinary work that each one would be assigned to. How spouses should forget about each other, how men and women should behave like brothers and sisters and not think of the world they had left behind.

Ted did not like rules and regulations in general but he understood the brothers-and-sisters recommendation. It was summertime, the setting was imposing and idyllic; all these men and women confined in close quarters could become an explosive situation. As a matter of fact, much later he arrived at a consensus with Aris and Mihali that these camps at Chandolin were endowed with a troubadour quality, namely, proneness to induce people to fall in love.

The first thing in the morning was a meditation session that lasted for three quarters of an hour. Then there was a copious breakfast. A short time later Ted found himself in the wood shop next to two men with serious faces. One was Antonio, the veteran Mexican Ted had met the day before. The other was Frederique, featuring a beard similar to Ted's. All three had been given the task of cleaning the darkened-by-time wooden boards from the railing of a chalet balcony. Each one was trying a different technique, sand paper, or scraping the wood with a broken glass or other sharp tool. Ted asked Frederique.

"What do you do in life?"

"I am orthodontist," said Frederique, "and you?"

"I am a physicist."

They both turned toward Antonio.

"I am chef," he said laconically.

"In a restaurant," Ted wanted to clarify.

"No, I am chef d'orchestre." He was an orchestra conductor!

And they continued the work they had been assigned. It wasn't long before Ted found out a strange thing about this kind of work. The aim was not to finish it and move on. The manual work provided an occupation while they would attempt the Work exercises Michel would give them during breakfast. Each day there would be a different exercise.

One day the exercise was: do not talk unless it is absolutely necessary. Ted did not think much of it in the beginning. But as the day advanced, he realized that there was practically nothing that would make it absolutely necessary to talk. So he ended up not talking at all that day.

The days went by and Ted made many friends. Despite a full schedule with almost no free time, the days were long and there were many occasions to meet people. Besides the three meals there were also two breaks—collations—one at 11 am and one at 5 pm. Following lunch a short siesta complemented the short-night sleep. After 5 pm there was one hour of "movements," oriental dances and physical exercises mostly designed by Gurdjieff himself. Then followed one hour of exchanges with the elders, periods of questions and answers in small groups often headed by Michel.

Ted was the only physicist and Michel made a note of this publicly at dinner one night. People seemed impressed but Ted felt rather uncomfortable. He was there searching for things that physics could not give him and yet they were honoring him for the very thing he found inadequate. He soon realized that it was Michel who unwillingly cultivated the veneration of physicists by his appreciation of scientific rigor. On many occasions Michel encouraged people to employ rigor of scientific caliber in their attitude toward the Work. Ted appreciated scientific rigor but did not see how it could be employed outside science. After all it was this very rigor that presented obstacles in his wanderings away from the paths of science.

Saturday was the camp's last day and work stopped early that afternoon. Men and women took saunas—separately of course—and received massages. In the evening there was a fete. Dinner was fancier and everyone dressed up. A small glass of Armagnac, Gurdjieff's favorite French cognac, was placed next to each plate. Toasts were made, jokes were told, but everything had a certain context appropriate to the setting. For example, Antonio told the following joke.

> In a monastery a young monk watches an older monk smoke while meditating. "How dare you?" he reprimands him.
> The older monk replies, "but I have permission from Father Superior. You can do the same if you only ask."
> A little later the young monk runs back distraught. "You played a trick on me. Father Superior was very upset that I asked such a thing."
> "What did you ask?"
> "I asked whether I could smoke while meditating."

"Wrong! You should have asked whether you could meditate while smoking."

Michel left the hall early after dinner and let the fete continue informally. Ted took the opportunity to impress the ladies with his knowledge of hypnosis. During the hypnosis course at Columbia he had learned how to carry out a profiling test for establishing how hypnotizable a subject is. So now he tried it on a willing volunteer, Marie-Louise, a sophisticated Parisian lady. She was a violinist; brunette, thin, tall, with green eyes, and she spoke rapidly as if always in a hurry; typically Parisian, Ted would say. But later he realized that in fact it was she who shaped his image of a typical Parisian lady.

A small group around him watched with excited anticipation.

"Do you know what you are doing?" an uptight character asked Ted with obvious concern.

"Just don't worry about it," Ted reassured him.

Ted gave Marie-Louise instructions in a slow suggestive voice, following the steps he remembered.

At some point he told her, "When I touch your left elbow, you will regain normal control of your left arm and you will retain only pleasant memories from this whole experience."

And proceeded, "Your left arm now feels lighter and lighter. It is being pulled up by an invisible string." Slowly but surely Marie-Louise's left arm rose involuntarily to a vertical position and remained there. When Ted forced it down, the arm rose again on its own.

But when the moment came, touching Marie-Louise's left elbow did not quite restore things to normal.

"My left arm feels funny," she kept complaining.

The uptight character was ready to call for help.

Ted told him to wait and repeated the whole hypnosis procedure from the beginning emphasizing the elbow cue. This time when he touched her elbow the arm felt normal again. Everyone calmed down.

At the end of the fete Ted and Marie-Louise went for a walk on the grassy hills around the chalet complex. She held nothing against him. On the contrary they felt close. They lay down watching the dark summer sky. With stars shining bright the sky seemed particularly close, as if it had come down lower.

"Even the dark space between stars seems bright," she said poetically.

"That's because it is very special. Have you ever wondered why it is dark," Ted asked mysteriously.

"What else could it be?"

"It could be ablaze with light."

"But there is no sun in that direction."

"Oh, yes there is. There is a star like our sun in *every* possible direction that your gaze may fall on the sky," said Ted emphatically, and continued, "the universe is so large and there are so many stars that no matter where you look there is a star. We just don't see its light."

"Because it is too far?" she guessed.

"Rather because light travels too slowly," he replied enigmatically. And then he went on, "The universe is more large than it is old, and before you tell me that I am comparing oranges to apples, let me explain. The universe is so large that there are distant stars in every possible direction. But most of these stars are so far away that their light has not yet reached us, despite light's extreme velocity. The light of most stars is still on the way to earth. On the other hand, some of the stars that we do see no longer exist but again the extinguishing of their light has not reached us yet. The picture of the sky we see does not represent reality at all. Some of the stars we see no longer exist and very many others that do exist we do not yet see. If light traveled instantly, there would be light everywhere in the sky. The dark space you see between stars is very special because it represents our only chance of actually looking at what was there *before* the universe was created."

They did not talk more. They lay there next to each other without touching. It was their last night in Chandolin and there was a lot to "digest" for both of them.

Coming down from the mountain the next morning Ted was jolted back into the real world: negotiating his way through highway traffic, seeing again his wife and children, taking shifts collecting data at the physics experiment at CERN, and running after entropy on Saturday mornings. Chandolin, Michel, and Marie-Louise all seemed like old dreams fading away but leaving emotional traces behind.

10 – Make Space for Time

Back at CERN Ted was quickly re-immersed in his familiar scientific environment. He found a certain comfort in this environment where objectivity reigns, emotional arguments have no place, and people's beliefs become irrelevant because science dictates the truth. A statement is either right or wrong, there is no point arguing about it; scientists have no choice but to take one position or the other. From this point of view, scientific work can enjoy a serenity akin to that stemming from esoteric bliss. Indeed, physicists walking around at CERN in their jeans and sweaters never seem to be angry, excited, or in a hurry. They display a calm similar to that Ted had observed among people at Chandolin.

Scientists generally seem void of emotions. They become excited only rarely, whenever they manage to make a discovery, or to be more precise, at the moment they make the announcement of their discovery. In that respect scientists display the same appreciation of the limelight as artists and performers.

But scientific discoveries are rare. Moreover, Ted's work involved the collaboration of many physicists, all of whom wanted to claim part of the credit for few minimum-impact results. There were no emotional rewards in store for him. He considered his work as factory research, and nostalgically remembered the days of his youth when he had been so fascinated with physics.

Ted now found more excitement in Gurdjieff's ideas. He suspected that there was more to be gained in Chandolin than he had received. He felt that when he was there he did not get access to all of it. There had been interesting people there and much food for thought but it had all happened in a short period of time. And he had been rather unprepared. He could not wait for his second chance next summer, but that would not be for eleven months. In the meantime he could do nothing but wait for the time to pass. He wondered how long eleven months would seem to him. Would he end up spending more or less than eleven calendar months on his own internal clock? Also, he liked living in the present. He had subscribed to the here-and-now maxim of the 1960s and never really let go of it.

The appreciation of time varies among individuals, disciplines, and cultures. Anthropologists have argued that the way a culture manages and thinks of time depends on the meaning its members find in life and the nature of human existence. Present-oriented cultures are relatively timeless and traditionless, and ignore the future. Past-oriented ones are mainly concerned with maintaining and restoring traditions in the present. Future-oriented cultures envisage a more desirable future and set out to realize it. The social concept of time involves two contrasting notions: time as a linear never-ending succession of minutes, hours, days, months, and years, and time as the circular reappearance of minutes, hours, days, weeks, and so on.

Ted had seen the results of a test that determined cultural approaches to time. People from different nationalities were asked to represent past, present, and future with three circles. The Belgians answered with three equal-size circles in a row meaning they considered present, past, and future equally important, with a very small overlap between present and past, but no overlap between present and future. The British had a significant overlap between present and past, but pictured the past as less important than the present and the future. The Germans pictured the future as most important, overlapping strongly with the present. The Russians pictured past and future more important than the present, but no circles touched; there was no connection between past, present, and future.

The Americans thought the future is much more important than the present and even more so compared to the past, all three circles lined up and touched. This was in contrast to the French, who found a significant overlap between past, present, and future with the past only little less important than the other two.

The Japanese answer was boldly different, three concentric circles!

Physicists do not think of time in these terms. They wouldn't approve of attributing unique importance to time. After all, time is only one of the world's dimensions, just as distance is another. In physics, time and space are coordinates of equal stature and an "equal-rights" argument could make the importance of space equal to that of time. The two dimensions represent two different worlds, and have their own "objects." Space objects correspond to the physical world around us, for example, a house or a car. Time objects correspond to events and happenings, for example, lightning or an earthquake.

There are patterns found in each world. Some patterns are orderly in space but disorderly in time. Snowflakes, for example, depict impressive order in space, with geometric patterns of high symmetry, but "disorder" over time; no two of them alike. Other patterns are orderly in time but disorderly in space. For example, highway driving involves small corrections with the steering wheel. In order to go straight, a correction to the right will always be followed by a correction to the left, and this symmetric pattern will repeat regularly over time. But there will be no repetitions in the ever-changing scenery around the vehicle's straight-line trajectory in space.

A much larger-scale example of the special relationship between time and space can be found in connection to the evolution of a developed country such as the UK through the centuries. In a snapshot of the state of development of different countries around the world today we can identify many intermediate states that the UK went through over the centuries. There are countries today where life resembles life in the UK one hundred years ago, or five hundred years ago. One can find communities in the world—for example, remote areas in the Amazon forest—where life today resembles life in the UK two thousand years ago.

Yet, from another point of view, even physicists may admit that time is more special than space because time can also enter our world in a more devious way, as an operator. Making an operator out of time consists of opening a time window and allowing a change to take place in it. The operation is called "derivation" in mathematical jargon, and its outcome is the rate of change. In simple words it translates to dividing the amount of change by the amount of time in which the change occurred. The ratio is large if the change is large, or if the time in which it takes place is short.

Ted had become aware of the importance of time as an operator in high school when he first learned that speed is distance divided by time, for example, miles per hour. The operation involved space and time, and resulted in speed. A car at sixty miles per hour makes it possible to reach a location sixty miles away in one hour. But in addition, speed turns out to be a variable that produces excitement to human beings. Experiencing speed is certainly more exciting than experiencing space or time directly. Moving quickly produces thrills, particularly to those who are unaccustomed to it. And this brings us to the double application of the time operator. For a person with no previous knowledge of speed, acquiring a strong taste of speed in a short period of time becomes an experience more intense than for a person who is already familiar with speed.

Ted had acquired a gut-level understanding of the powerful consequences time had as an operator and he instinctively applied it in many social situations.

11 – Intricate Aspects of Time

There was excitement in the Gurdjieff group on the weekends that Michel de Salzmann came to Geneva. Among other things they organized a Saturday night dinner in the tradition of Chandolin. Ted looked forward to theses occasions, because dinner lasted for a long time and Michel directed a free-form exchange between the elders and other group members. But no matter how many people were present at dinner there was always only one conversation.

On one Saturday-night dinner the topic of discussion was photographs. Michel triggered the discussion by asking in his rhetorical style, "Why are photographs interesting? Why do people like going to museums?"

The silence that usually followed Michel's questions gave Ted a chance to prepare his response, and before anyone else spoke he said, "Photographs, like museum exhibits, present an interest only as a new experience. One rarely visits a museum to see the same exhibit for a second time. The same thing is true with photographs. Most of the pleasure from photographs comes when one first looks at them, or, years later, when all is forgotten and it feels as if one is looking at them for the first time. New impressions are nourishing."

Ted knew well that everyone there was familiar with this idea. In his book *In Search of the Miraculous* Ouspensky argued that new impressions play an important nourishing role and can be considered as an indispensable ingredient for life. Ouspensky classified fundamental needs according to how much we depend on them. He considered ordinary food as the least significant need because one can survive up to a month without it. Water is a more fundamental need because without it we can live only for a week. A still more essential need is air (oxygen), without which we die in a few minutes. Ultimately, the most refined and most essential "food" of all is impressions composed of a variety of external stimuli. Ouspensky claimed that one might die in a matter of seconds if deprived completely of all sensory impressions.

It is difficult to verify practically the validity of such a hypothesis. There is some evidence, however, which points in that direction. In sensory deprivation experiments subjects wore gloves and were left to float on water in a dark, soundproof cells. In a matter of hours they entered a temporary catatonic state, a condition similar to being unconscious, half-dead.

Further proof may lie in the fact that repeated identical stimulation (lack of variety) deadens sensitivity. Pickpockets put this into practice when they find ways to stimulate the wallet region of their victims before they pull the wallet out. Soon after putting on perfume, you no longer smell it because you have become accustomed to it. The iris of the eye oscillates rapidly and continuously to avoid stimulating the same cells on the retina. Looking at a uniformly blue sky can produce a sensation of blindness. The same effect is obtained if you cut a ping-pong ball in half and cover your eyes so that all visual stimuli become undifferentiated.

Finally, it is common knowledge how easily young children become bored. In urgent need to be nourished and grow, children seek new impressions ravenously. The more possibilities a toy offers, the longer they will spend with it, but soon will turn elsewhere for more new impressions. Older people, on the other hand, do not search quite so insatiably for new impressions. Many prefer more meditative activities and a state of mental tranquility rather than continual sensory stimulation. For an older person, an impression may no longer be new. Nevertheless, the person must still be nourished to stay alive and in good health, so that a need for some form of sensory stimulation and new impressions continues throughout life.

"The pursuit of new impressions," continued Ted, "is the active agent behind visits to museums, movie-going, and tourism in general. It is new impressions for which people travel, sometimes to improve their health on their doctor's recommendation. Wise schools of thought, often from Eastern cultures, teach techniques for looking at familiar things in *new* ways."

Michel nodded in approval with Ted's answer. But José, Michel's wife, was upset. "Do you have a wife, Sir?" she thrashed at him for advocating the merits of diversity.

Ted was not intimidated. He understood how impressions work on a fundamental level. He knew that the richness and intensity of new impressions is related to the amount of variation *per unit of time*. He had gone through a similar realization years earlier while pondering on the way life feeds on change, and the way routine and stability kill sensitivity and enthusiasm. It is true that the euphoric effect of a salary increase is rather ephemeral. Soon after a salary increase the employee will come back asking for a new one.

Abrupt transitions generate sensations. The larger the separation between the two levels and the faster the transition the more intense the sensation associated with the jump. This is what happens when we "operate" with time.

Moving in space with a certain speed—a simple operation by time—is exciting because scenery and visual impressions change continuously and rapidly. But moving for the first time, is a double operation by time, and consequently even more exciting, because besides new scenery there is the new experience of perception in motion.

Speed, of course, can eventually become boring. Constant speed amounts to what physicists call an inertial frame, that is, an environment in which with eyes closed one can no longer detect the motion. This becomes evident in an airplane where visual stimulus is limited. But even on the highway, driving on a straight section at a constant speed makes one unaware of the speed magnitude. In a somewhat self-defeating move highway designers have sometimes introduced unnecessary turns for the sole purpose to keep drivers awake.

An alternative to introducing bends in the roads would be to introduce variations in the speed, that is, to apply a second operation by time. Try the following exercise next time you find yourself on the highway. Maintain a constant speed of sixty miles per hour for at least five minutes. Then drop your speed rapidly—and carefully—to forty and keep it stable there for another five minutes, following which, you raise it again as fast as you can to sixty. The experience offers surprisingly sharp sensations. Right after dropping the speed, and for only a short period of time, the speed seems absurdly low; similarly the high speed initially seems excessively high. Switching between speeds like this you can maintain a fifty-mile per hour average speed and stay wide-awake, even if the road has no turns.

Operating by time repeatedly is becoming the preferred eyepiece through which one looks at the increasingly demanding challenges of today's society. Industrial success clings to growth, namely making more revenue this year than last year—first operation by time. But often CEO's goal becomes not only positive growth, but also positive rate of growth, that is, surpassing last year by more than last year surpassed the year before—a double operation by time.

The interest in a double operation by time is understandable. While the first time derivative produces speed, the second time derivative produces acceleration, which translates directly into force. It does not suffice to move. One further needs to be forceful. You feel a car's acceleration, or deceleration, in your guts even with your eyes closed. These are more intense sensations than the simple fascination associated with speed. In air travel thrill-hungry passengers and youngsters find more excitement in the forceful take-off and landing moments.

Beside space, time can operate in a similar way on many other variables. Let us consider action, for example. Action is a variable rigorously defined by physicists to reflect in a systematic way what we mean by the everyday use of this word. It is equal to the product of the distance and the speed with which the distance is covered (or, alternatively defined, the product of the amount of energy spent and the amount of time required to spend it.) Action in *natural* phenomena is always minimized. Whatever happens in nature, e.g., the way light travels, the trajectory of a rock you may throw, and the way water falls down a waterfall, are always characterized by the least action possible (this is more than we can say for human endeavors; people's actions are generally wasteful in energy and time.)

Operating by time on action repeatedly—as we did on space earlier—produces first energy and then power. The social analogy is obvious. The actions of an individual can be measured by how fast he or she goes how far (or, alternatively, how much energy he or she spends how fast.) Dividing by the time in which this is done gives a measure of the individual's energy. Finally the amount of energy *per time* is a measure of the individual's power.

Many company executives welcome traveling because it makes them look active. The faster they move and the further they go, the more active they appear to be. If you want to know how energetic an individual or a working group is, you count the actions they accomplish over a period of time. If you want to know how powerful they are, you measure how their energy output changes over time. One's power diminishes if he or she accomplishes fewer or lesser actions per unit of time.

The second time derivative can also demystify intuition, which at times has been attributed to supernatural origins; it may not be supernatural at all. Intuition results from sensitivity to the forces of the second derivative. Successful stock analysts use their gut feeling more than their knowledge in anticipating future stock performance. For a stock price, the second derivative corresponds to the stock's rating (AAA, AA-, etc). A stock's return of investment (the first derivative) over a certain period of time is a less reliable predictor of the stock's future performance than the stock's rating.

For a company, the second derivative corresponds to the prospects for growth. The feelings of employees or stockholders may carry a more realistic assessment of the company's future than the opinion of its management team. Managers are biased by the expectations stemming from their strategies, their plans for action, and their targets. They extrapolate past growth rates as if they were in a vehicle moving with constant speed. But low-level employees and stockholders use only their gut feeling to rate the company's future prospects, their optimism or pessimism, which like the acceleration in a vehicle is evidenced via a force (second derivative).

12 – The Humanities of Science

As time passed Ted watched his attachment to the scientific work he was doing at CERN erode. At the same time he became more sensitive to scientific "irregularities," incidents of inappropriate behavior by physicists that he referred to as "The Humanities of Science." His collection of them kept increasing. The word humanities in this case had a bad connotation. Physicists are not supposed to succumb to human weaknesses, such as lying, having biased opinions or preferences, and needing others' approval.

There was the story of a physicist going to court in order to claim a Nobel Prize. Ted found it bizarre not only because a Nobel Prize is a distinction offered, not asked for—much less legally demanded—but also because the lawsuit came twenty years too late!

Oreste Piccioni, an Italian-born experimental particle physicist, felt he had been deprived of recognition for his contributions to the discovery of the antiproton in 1955—the first known particle of antimatter. In 1954, Piccioni had visited the University of California at Berkeley and discussed with colleagues Emilio Segrè and Owen Chamberlain how magnets might be used to guide a beam of antiprotons to instruments that could detect them. The next year, Segrè and Chamberlain discovered the antiproton. They did not think Piccioni's suggestion was central to what they did and did not list him as an author on the paper announcing the discovery. Piccioni felt he had provided the blueprint for the experiment. He complained to colleagues for years, especially after Segrè and Chamberlain received the Nobel Prize in Physics in 1959. In 1972, Piccioni filed a lawsuit against Segrè and Chamberlain seeking $125,000 in damages and a public acknowledgment of his contributions. The courts dismissed the lawsuit, saying it had been filed too late.

A more serious case of "humanities of science" was the evolution of the Michel parameter, which Ted had studied in graduate school. In 1949 the French theoretical physicist Louis Michel had defined a parameter whose value was constraint between 0 and 1, and which would have important consequences on our understanding of the weak nuclear force. Immediately experiments were set up to measure this parameter.

The first measurement in 1949 determined the parameter to be equal to 0 within the experimental error. A second experiment in 1951 confirmed the previous result with increased precision (smaller experimental errors). A little later a small but non-zero value was reported, but compatible with the previous two measurements within the experimental errors. This game continued for a decade. Every new measurement would determine a slightly higher value for the Michel parameter, but always compatible with the previous determination within the experimental errors. It must be pointed out that theoretical work advanced in parallel and concluded in 1957 that the Michel parameter must be equal to 0.75. Well, this is the experimental value that was finally measured in the early 1960s with high precision.

It seemed as if a constant of nature had drifted over a period of ten years in tune with the shifting of its theoretical prediction. And yet, no *official* mistakes had been made. Every experimental result at the time of its publication was in rigorous scientific agreement with the most recently published result.

It was a case of legitimized insincerity. When the theorists predicted zero in the beginning, the experimentalists measured zero. When the theorists finally predicted 0.75, the experimentalists measured 0.75. In the process they made sure not to call their colleagues liars. At each measurement they biased their results—be it subconsciously—within the limits permitted by the experimental error. This did not make Ted proud of being an experimental physicist. He had always considered experiment as the basis for scientific research. He liked to summarize the entire scientific method in three words beginning and ending with experimentation:

$$\text{Observation} \rightarrow \text{Theory} \rightarrow \text{Verification}$$

Another Italian-born physicist, Paolo Franzini, also left a dark mark in the history of physics. He was one of Ted's teachers at Columbia University. Franzini had carried out an experiment that purportedly observed a violation of the symmetry in particle-antiparticle conjugation. Such an observation would have important theoretical consequences. Therefore journalists were mobilized and articles appeared in *The New York Times* and other popular press. But the experimental result was wrong and it was proven as such three years later through a more precise follow-up measurement by another Columbia physicist, Wonyong Lee.

Franzini, who did not believe his own result, had tried to understate his "discovery" at the time of publication by the inclusion of the following statement:

> "...our observed asymmetry admits the possibility of a C violation as large as the theoretical maximum. Since the asymmetry differs from zero by only two standard deviations, we can reach no definite conclusion."

However, before proven wrong, this experimental result made headlines across the world of particle physics and played a crucial role in Franzini's award of tenure at Columbia.

There was also the case of German physicist Heinz Filthuth, head of the High Energy Institute at the University of Heidelberg. He liked physics but also good life. He spent the Institute's budget on both. Among other things he bought two identical luxury cars, a high-end Mercedes model, and kept one at CERN in Geneva and another one in Heidelberg. Everyone thought he had only one and drove it back and forth. But he flew instead. He was eventually caught and disgraced, even though he personally believed that he had done nothing wrong.

And then came the *psi*chology affair. In 1974 American physicist Burton Richter working at the Stanford Linear Accelerator, discovered what he called the Ψ (psi) particle. At the same time another experimental team working at Brookhaven's Alternating Gradient Synchrotron under Chinese-born Samuel Ting reported the discovery of what they called the J particle. The two particles turned out to be one and the same, and its discovery helped confirm the existence of the *charmed quark* (by now physicists had abandoned old traditions and were adopting names irrespective of the word's meaning in everyday usage—sign of decline). The two physicists shared the Nobel Prize in 1976 and the event went down in history as a simultaneous independent discovery of the J/Ψ particle. But some physicists at CERN believed that Richter could not have made his discovery without some inside information secretly obtained from Ting's experiment.

The two experiments approached the physics question from opposite directions. Ting studied the production of particle-antiparticle pairs (in this case electron-positron pairs) and from their characteristics tried to infer that a larger particle—in this case the J/Ψ—must have decayed and produced the pair. In contrast, Richter had at his disposal a special particle accelerator, a linear one, consisting of two colliding beams, one with electrons and one with positrons. He could thus evidence the existence of the massive particle by first producing it and then observe its decay. But whereas Ting observed pairs with a whole spectrum of energies coming from many sources among which was the decay of the J/Ψ, Richter had to set the energies of the electron/positron beams to the right value in order to produce and then observe only this particle. Each beam should have electrons (and positrons) with energy equal to exactly half the mass of the J/Ψ. But not knowing the particle's mass yet, Richter had his hands tied. The setting of the energies had to be precise otherwise he would be looking for a needle in a haystack. Rough sweeping scans over the entire energy spectrum could help only if they were gifted with extreme luck.

Ting's experiment had been running for a couple of years. For a period of six months his team kept the discovery as a top secret while the team tried to understand it and eliminated all possibilities of error. Contrary to Richter's approach, the method Ting used to evidence the existence of a particle was messy and required elaborate data manipulation. Over the six-month period, some rumors leaked outside Ting's group, how they had discovered an important new particle. But the group's solidarity stood firm behind Ting's instructions and no parameters of the new particle were leaked. They all knew that Richter could "see" this particle within hours of the time that he knew at what value to set the beam energies of his linear accelerator.

Therefore and despite lack of concrete proof many physicists at CERN speculated that Richter could have dispatched "spies" to Brookhaven once rumors about the new particle reached his lab. Physicists can make good spies and it wouldn't be long before they'd pinpoint the mass of the phantom particle via information concerning demands Ting made on the support staff of the Brookhaven Lab, such as current values for his experiment's magnets.

At about the time Ting was ready to publish his findings, he visited Stanford University for a routine meeting of scientists associated with the Stanford Linear Accelerator Center. At that meeting, he announced that he was about to publish a major experimental result on a new particle. To his amazement he heard Richter saying that he had just demonstrated the existence of a similar particle, which he had named the "psi" particle. Ting rushed back and within 24 hours had submitted an article for publication. He just made it. Richter's article had already been submitted. The two articles appeared in the same issue of *Physics Letters*.

Bitterness lingered among Ting sympathizers for some time. Richter was now able to produce the J/Ψ easily and in large quantities. He studied its properties rapidly and exhaustively. His group published articles one after another on the subject. Two months later there was a presentation of their results at CERN but not by Richter himself. A young French physicist was delegated to report on what Richter's group now referred to as "psichology," all one would ever want to know about the Ψ particle. The presentation drew a large crowd that filled the CERN auditorium. At the end during the question-and-answer period Jack Steinberger had a question, which he put in a loud clear voice from the last row of the amphitheater.

"Can you tell us how often the J particle decays into two muons?"

The speaker put on an acknowledging smile at Jack's reference to the particle with Ting's name and proceeded to answer the question.

Ted had followed all these developments as a spectator interested more in the humanities of science than in the physics itself. But two years later

Richter, a fresh Nobel laureate, came to CERN and joined Ted's experimental group. He wanted to add a new apparatus on top of the existing experimental setup in order to detect muon particles and thus evidence in a yet another way the charmed quark already proven via "his" Ψ particle discovery.

Hosting a Nobel laureate added certain excitement in Ted's group, if not funds and support. Richter submitted a proposal for his experiment and before approval was granted by the CERN committee Ted faced the daunting task of presenting and defending Richter's proposal to a select group of physicists that included Carlo Rubbia. Ted was nervous and did not really want to do it, but Richter was traveling and there were no other physicists around from the group to do the job.

Richter's recent appearance and noisy presence at CERN had not gone down well with Rubbia, who had not yet received a Nobel Prize. Before Ted had a chance to finish presenting his transparencies, Rubbia raised a number of objections.

"How do you intend to detect the muon particles?"

"By inserting a 60 centimeter-thick slab of iron to stop all other particles."

"60 centimeters of iron is not enough. Iron's radiation length is only 1,76 cm and 60 centimeters will not give you clean muon signatures on the other side. This experiment will never work!"

Ted had done his homework and could give the right answers. But it wasn't a question of science. The handful of physicists around Rubbia kept nodding their heads in approval as Rubbia kept criticizing.

Ted was demoralized.

"This is not my proposal," he finally explained, "Burton Richter has made these calculations and he believes there will be a clean muon signal on the other side of the iron slab."

Rubbia laughed derisively. Ted sat down with a bad taste in his mouth. He later recounted the whole incident to Stoyan, a Bulgarian physicist friend.

"I should have challenged Rubbia to a bet," said Ted bitterly.

"It wouldn't have helped," replied Stoyan. "It is not what you say that counts, but who you are, even in science."

Stoyan had a streetwise attitude about many things in life. Once he taught Ted how to rejuvenate an aging car battery: you simply charge it fully and then short its poles, several times in rapid succession. On another occasion he had summarized the quintessence of World War II in a nutshell: it changed a warlike people (Germans) to businesslike, and a businesslike people (Jews) to warlike. Concerning Ted's frustrating confrontation with Rubbia Stoyan's verdict was, "You simply were not in the league."

13 – Accidents, Karma, Destiny, Miracles

That winter Ted had a reunion with Aris and Mihali in Athens. He found out that after they had separated in New York City Aris continued to frequent the Gurdjieff Institute there. More than that, he rapidly climbed up the hierarchical ladder and four years later, when he moved back to Greece, he was in a position to found his own esoteric school along the lines of Gurdjieff's teachings there. He had become a spiritual guru to some twenty-five Athenians. Mihali was also visiting Athens frequently. He had gotten a job in Saudi Arabia and "commuted" between Athens and Jeddah often.

When the three of them met again after such a long time there was much to catch up with. Ted did most of the talking, recounting details from Chandolin and Michel de Salzmann. Aris was practically envious. His contacts with the Gurdjieff Foundation were in far-away America and with people of lesser stature than Michel de Salzmann. The three of them day-dreamed of attending a camp at Chandolin together, not a trivial matter considering how long Ted had to wait before being admitted. He suggested that Aris and Mihali write to Michel asking for permission. Ted then would put in a word with Michel in their favor.

Around May the following year and before the activities at the Gurdjieff group stopped for the summer recess, Michel de Salzmann and Ted met in private. Michel had occasional tête-à-tête meetings with group members to give custom-made instruction and obtain feedback on how they advanced with the "Work," which is how people at the Gurdjieff group referred to the teachings and the practices that enable one's development toward higher states of consciousness. This time he told Ted that in anticipation of the summer camp Ted should prepare a talk on "Physics and the Work" and deliver it one evening after dinner at Chandolin. "I would like to see how a physicist sees what we are doing here," concluded Michel.

It felt as if Ted would not be able to distance himself from science even during his retreat at Chandolin. But strangely, the prospect of using physics in an unorthodox environment appealed to him. He knew exactly what he was going to say. It was a topic he first came across under stress while taking his qualifying exams before beginning the research for his Ph.D. thesis. Despite the seriousness of the situation, one of the exam questions had made Ted pause and drift into philosophical contemplations.

It was the third day of a series of examinations designed to qualify graduate physics students as competent to carry out research towards a Ph.D. The marathon examinations covered all of physics, or one could say, all of everything because there was no limit to the kind of question that may be asked. There had been such questions as what is the average density of the rocks in Central Park, and at what temperature do we fry eggs.

The particular question that triggered Ted's excursion into philosophy was to compare the ratio of the diameter to the distance to closest neighbor, of the following: the atoms in a piece of charcoal, the molecules of an ideal gas, the planets in our solar system, the suns in a galaxy, and the galaxies in the universe.

As above so below, had thought Ted. Not long before, he had seen a short film on this subject. The only action in this movie is the camera zoom, no words, no sound. Starting from a boy and his father in a boat, the camera zooms in onto the boy's arm where a mosquito sits. We see the mosquito's trunk as a huge cylinder disappearing under the rough surface, a close-up of the skin. We enter the reddish material and become progressively aware of its granular structure. Soon we begin distinguishing the cells of the skin, each with a dark nucleus. As the zooming process continues, we become able to distinguish the molecular structure of a cell, and before long a lot of free space appears between enlarged molecules. The next phase brings us inside the molecule where the atoms are now visible. Zeroing in on one particular atom we penetrate the electron cloud only to realize that there is nothing inside. Almost nothing, a tiny little dot in its center the nucleus; all the rest is empty space. Soon after we zoom in onto the nucleus with its protons and neutrons, we are able to distinguish their internal structure of three constituent bodies, the quarks.

Suddenly the zooming stops and its direction is reversed, now backing out of the proton, out of the atom, out of the molecule and the cell to find the mosquito on the arm of the boy. But the action does not stop here. The zooming continues backwards. Receding away from the boat we realize that the boat is on a lake that is flanked with mountains. From a larger distance away we get an airplane view and later the curvature of the earth becomes apparent. As we recede from the earth the moon comes into the field of vision and soon other planets around it. At some point, the full solar system fills the screen only to shrink down to a dot and together with many other such dots (suns) outline our galaxy, the Milky Way, first as a disk that slowly shrinks, later as a dot with other galaxies also coming in as dots. Then the zooming stops once again and reverses direction to penetrate successively, galaxies, the Milky Way, the Solar System, the Earth's atmosphere, and come back down in the boat with the boy, his father, and the mosquito.

There is no message spelled out in this short film, but there emerges a pronounced resemblance between the worlds above us (stars and galaxies) and the worlds below us (molecules and atoms). The structure of the universe appears as a grand continuum made of nested patterns, which repeat themselves. We humans (the boy and his dad) are innocuously situated somewhere in the middle.

For his Chandolin talk Ted was going to present the different worlds nested like Russian dolls with the human world positioned in the middle of the range. He could do all this not only quantitatively, as opposed to just artistically as in the short film, but also from a much bigger-picture point of view. He had found impressive facts in a book by Rodney Collin called *The Theory of Celestial Influence*.

Collin was a physicist who approached many of Gurdjieff's ideas from a scientific angle. For example, Collin clarified the often-confused notions of accident, karma, destiny, and miracle. Physics has room only for the first one—accidents; the last three have no place in the hard sciences. And yet Collin gave scientific definitions for all four of them. The definitions depended on the timing of the cause-and-effect relationship in each case. In a pure accident the cause immediately precedes the effect. For example, you come hurriedly out from the entrance of a store and you bump into a passerby. The accident's cause goes back only a few seconds to the time you decided to move. But if the cause is due to repetitions or habit, then it is karma. Example, you get caught speeding despite carefully watching out for radar traps. You may have gotten away with speeding many times, but eventually you will be caught because of your karma, the constant repetition you indulge in. In destiny the cause goes far back, before you were born. Example, you are an African American because your parents were African American; you had nothing to do with it. And finally, in a miracle the cause lies neither at the present, nor in the repetition, nor in the far past, but could come anytime including *after* the effect, as would be the case if you succeeded in locking a drawer after having dropped the key inside!

In that book Ted had read that there is a factor of about 28,000 in the time-frames between adjacent worlds in the nested-world model he wanted to present. To prepare his talk he went back to the book and dug into the details. It was like doing physics but now with an all-permitting "poetic" license. He manipulated numbers and drew conclusions that lead to fascinating revelations. The nested-worlds model was not perfect and the factor 28,000 did not always do the job, but there was enough symmetry to make pure coincidence highly unlikely. He felt that he was making discoveries. It gave him one more reason to be excited about participating in the upcoming Chandolin camp.

14 – As Above so Below

In his talk at Chandolin Ted presented a big picture of the universe, much bigger than physicists normally admit. He was aware that the work of Rodney Collin was little known, much less understood, by those in the audience despite the fact that it constituted a scientific approach to Gurdjieff's teachings. It was precisely because Collin tried to put science to Gurdjieff's teachings that Gurdjieff's followers stayed away from Collin's work. For them, a scientific point of view precluded widespread understanding. Ted felt a familiar frustration: people's mental block against physics and science in general as being too difficult to understand. But he was determined to persevere.

He began his talk with, "The theme of my talk is 'As above so below'." He knew that the people there were familiar with this notion for reasons different from his own.

"The universe consists of many worlds that are nested like Russian dolls," Ted went on. "There are worlds larger than ours such as the biosphere often referred to as Gaia, the solar system, and the galaxies beyond, as well as worlds smaller than ours: the cells in our body, the molecules of these cells, and even smaller entities.

"Rodney Collin argues that each world constitutes a cross section, a projection with one dimension less than the next world up. For example, the world of a blood cell is a cross section of the artery where the cell happens to be at a given moment. Another cross section higher up the artery, where the cell will be at a later moment, represents a 'future' world for the cell. When the cell is at the heart it has no knowledge of the existence of the brain. The various organs of the human body exist for the cell only *in time*. Thus the *third* dimension (length for humans) corresponds to the *fourth* dimension (time for the blood cell.) For a molecule, however, moving within the cell it would be the cell's third dimension that would represent time, whereas the cell's time (or man's third dimension) would correspond to a fifth dimension, something outside the molecule's time dimension. This fifth dimension is an unfamiliar concept mysteriously connected to the idea of survival after death or the possibility of reincarnation or repeated existence.

"From the electron's point of view, whose time dimension echoes the third dimension of the molecule, man's third dimension is difficult to comprehend, because it represent electron's sixth dimension corresponding to all possibilities of repeated existence, some kind of an all-encompassing limit.

"In this model each world needs six dimensions to be completely described. The first three dimensions describe space, the fourth is time, the fifth is all lifetimes (all repetitions in time) tantamount to eternity, and the sixth is all possibilities of existence including unimaginable ones.

"This 6-dimensional view of the universe yields interesting insights. Looking with zero dimensions—as points—all worlds appear identical. Similarly looking with six dimensions—encompassing all possibilities—all worlds again appear identical. However, looking with one to five dimensions, the worlds appear progressively first more different and then more similar.

"For example, an organism that is ten times longer than another organism in one dimension will appear one hundred times bigger in two dimensions. In three dimensions—seen as a solid—will have one thousand times more volume. As we add dimensions one at a time up to three, the two organisms appear more and more different, more separate, more clearly distinguishable, from each other.

"But introducing the fourth dimension—the notion of lifetime—makes similarities emerge. All living organisms large or small trace the same cycle in time: conception, birth, maturity, and death. In our 3-dimensional world a fly and an elephant have nothing in common, but in four dimensions, that is including their life cycle, common characteristics begin to emerge. Even more diverse entities will show things in common in five dimensions, that is including repetitions of cycles around a vital center. Blood cells, humans, moons, and planets reveal significant similarities.

"There is no doubt that maximum differentiation occurs in three dimensions. Living in a 3-dimensional world makes us experience existence in its coldest, most separate, and most exclusive aspect. This may explain the much-talked-about loneliness and desolation of contemporary men and women in their 3-dimensional view of the world.

"In a 3-dimensional world there is height, breadth, and width. But different worlds do not share the same dimensions. Think of a brick and a house, each is a 3-dimensional object, i.e., it has its height, breadth, and width. But whereas one brick is 3-dimensional, many bricks in a straight line can be approximated from the house's point of view as a 1-dimensional object, with length its only dimension.

"Similarly for human beings, they are 3-dimensional entities living and moving around in their 3-dimensional world. But seen from the earth's point of view, the human world is largely confined to a 2-dimensional surface, the surface of the earth. The third dimension of the human world has practically nothing to do with earth's third dimension that extends from its molten core to the moon the sun and beyond.

"So in this celestial hierarchy, each world appears to lack the lowest dimension of the world below but features a new dimension above and beyond the reach of that world. Every world still has three special dimensions, but shares only two of these dimensions with the adjacent worlds. It possesses a new dimension that the world below lacks and lacks one of the dimensions of the world above. This means that each world is partly invisible to worlds greater and smaller than itself. The further a celestial world is from us the more of it becomes invisible to us.

"A cell, as revealed by a microscope, may be a 3-dimensional organism, but its world lies three world's below the human world so to humans the cell seems as an immeasurable point practically dimensionless. The situation of our solar system within the Milky Way is similar to that of a single blood cell within the human body. A white corpuscle is also composed of a nucleus (sun), with its cytoplasm (sphere of influence), and it too is surrounded on all sides by millions of similar and other cells, the whole of which forms a great being.

"But what kind of being is our sun? What does it really look like? To answer this question we must consider the timeframes appropriate for each word.

"Human beings are characterized by four cosmic timeframes, each corresponding to the period of time required to 'digest' some primary kind of nourishment. The day-night cycle—24 hours—is the time required to complete a cycle of digestion of solid food. The respiration cycle—around 3 seconds—is the time required to 'digest' air (oxygen), a more refined and more essential type of food. There is an even more refined type of food, light, the perception of which (or 'digestion' of) requires less than one thousandth of a second. Finally, a typical lifetime of 80 years corresponds to a period of digestion of all experience in one's life.

"There is a curious symmetry that permeates these four timeframes. There is a factor of 28,000 between them. An average human life of 80 years contains around 28,000 days and each day sees about 28,000 human respirations of 3 seconds each. One more factor of 28,000 down brings us to the level of the limit of our perception, i.e., we cannot perceive a flash of light whose duration is 3/28,000 seconds or less.

"But a day in one world is as long as a lifetime in the world below, and at the same time a respiration in the world above, and so on. The old saying has it that the life of a man is but a day for nature. Similarly the day of a man is almost a lifetime for cells in his body, and a day of a cell is a lifetime for its molecules. Indeed if we define a blood-cell's 'day' as 30 seconds—the time it takes to complete its work cycle making the round trip heart-tissue-lungs-heart—it corresponds to a lifetime for the cell's constituent molecules that get destroyed and rearranged in the process. Each time a blood cell passes from the lungs and becomes oxygenated its molecules 'die' and are 'reborn.'

"In order to observe something correctly the observation must be made in the object's proper world, that is, the world in which the object enjoys its own three space dimensions. To observe the sun in this context we must look at it from its world that lays three worlds higher than ours. This implies that our unit of perception must increase by a factor of 28,000 x 28,000 x 28,000 = 22 trillion. Our usual unit of perception of 3/28,000 seconds must then become 80 years. With this new unit of perception a 'glance' at the sun reveals quite a different entity.

"The sun, while being the center of our solar system, is moving at about 12 miles a second toward another star, Vega. In 80 years the sun will have traced 30 billion miles into space carrying the entire solar system with it. The diameter of Neptune's (our outermost planet) orbit is about 5.6 billion miles. Consequently the body of our solar system in 80 years shows up as a figure about five times longer than it is wide, or roughly the proportion of a human figure.

"Furthermore, one expects different colored sheaths to envelope the white-hot central thread of the sun. These sheaths are formed by the manifold spirals of planet trajectories (with frills woven around them by their moon trajectories), as well as by other eccentric trajectories of myriad asteroids and comets. These wrappings around the sun's elongated figure add the final touches. The figure that finally emerges is rather imposing. Could it also possess life and consciousness?

"If we look directly at the sun in the sky we are unable to describe what we see. If we persist looking at it, we'll become blind. Why is the sun so different from everything else in our world? Why are we not able to look at it?

"The world of the sun is three worlds higher than ours; it consequently shares no spatial dimensions with our world. Moreover, its third dimension corresponds to our sixth dimension, i.e., all possibilities for humans. When we look at the sun we are looking through a hole in our 3-dimensional environment, out into the 6-dimensional world."

Ted arrived at the end of his presentation. He wanted to finish with a punch line.

"Has any one of you seen the light?" he asked before sitting down.

It was a loaded and provocative question. Loaded because it was the archetypical question asked by hippies back in the 1960s while tripping on LSD. Provocative because there were elders in the assembly, Michel first among them, who in principle had achieved a state of development capable of perceiving much more than a novice such as Ted. Could they look at the sun without burning their eyes?

A silence followed that seemed to last forever. It became embarrassing and Ted felt the need to defuse it. He began talking again, without standing up.

"Man's perennial attribution of deity to the sun can now be better understood. Besides its life-sustaining function the sun demands respect and punishes insolence. The mechanisms involved in the splitting of the atom and the fabrication of the hydrogen bombs are the very mechanisms by which the sun gets its own nourishment. They don't belong to the beings of three worlds below, not before these beings evolve to a comparable level of consciousness. Proof is that after making bombs for killing each other, human beings became intoxicated by nuclear energy and began using it rampantly, leading to more deaths via a series of accidents. The opposition expressed by the environmentalists was perhaps more deeply rooted than they thought."

Ted's last remarks broke the silence and several exchanges followed but a sober atmosphere lingered until the end of the evening. Afterward, neither Michel nor anyone else gave Ted specific feedback on what he talked about. In this crowd, compliments, approbation, and even good manners were considered unnecessary and artificial. But Ted felt satisfied and full of energy. He knew that he had brought valuable pieces of knowledge to an audience that should appreciate it.

15 – Mathematics with Social Use

At the CERN program inquiry office, where physicists went for help with their complicated software programs, worked several experts answering programming questions. One of them, Bjoern, had made an unpleasant impression on Ted because of his arrogant attitude. Bjoern generally received Ted coldly and basically treated him like an idiot who did not know how to write computer programs. Ted had been taken aback at first with Bjoern's conceit. At CERN physicists were supposed to be the kings. Everyone else was there to simply serve them. But with Ted's frequent visits the two men began to know each other better and eventually became friends.

Ted understood the origin of Bjoern's arrogance as stemming from his eccentric genius. He was a brilliant mathematician who combined several other incongruous behaviors. He was Swedish, a devout Catholic—which is rare for Swedes—a regular gambler, a father of nine children by three different women two of whom had been his wives. In fact, at one point he had lived for more than a year with two women simultaneously under the same roof. One was his wife; the other one his companion. They even had a common bank account with all three names.

"Which one would you go to bed with at night," Ted had asked incredulously.

"Well, it depended," Bjoern answered in his usual nonchalant way. "We would all three sit on the couch in the living room watching TV and at some point two of us would withdraw to the bedroom." All this had not lasted much more than one year. Bjoern became estranged from both those women, and ended up marrying a third one. But things were now faltering once again.

Ted and Bjoern began seeing each other outside CERN. Bjoern was smarter and more scientifically minded than most physicists. Ted would solicit Bjoern's opinion on the widest range of issues. One day Ted asked him, "What makes a woman attractive?" and he elaborated.

"We have been brainwashed by television and the media to consider women attractive when they conform to stereotypes: young, blond, thin, blue eyes, big breasts, etc. But how real is that? Why can't an older woman be sexually attractive? And when I say older, is there an age limit? Once under the influence of marijuana I began tearing down all stereotypes imposed on us by the outside world. I began questioning whether man's sexuality had anything to do with women at all. Why not with other men, or animals, or inanimate things for that matter. Where does one draw the line? Can a chair be a sexually attractive object? Are there any guidelines?"

Ted had raised the same question with Mihali back in New York City. Mihali's reaction had been, "Watch out, my friend, when you make *tabula rasa,* not to tear down all points of reference and become lost!"

Bjoern did not think long before answering.

"It is simple," he said. "For a woman to be sexually attractive she must be of childbearing age."

Ted did not react right away. This kind of answer, a stereotype on its own merit, was what he would have expected from Bible preachers and door-to-door Christian proselytizers. Ted always closed his door to them. But this time it was different because the message was coming from Bjoern. (Does banality become profound when it is pronounced by a brilliant scientist?)

Bjoern sensed Ted's dilemma.

"It may sound as an overly simple answer, but you physicists argue that the simpler the law, the more universal its validity. Just take it as a light in the dark to prevent you from mistaking chair legs for sex objects," he snapped.

The casino Bjoern frequented was in Divonne, France, not far from Geneva. It had long tradition but in contrast to its flamboyant rival in Monte Carlo, it cultivated discretion. Responding to Ted's request Bjoern invited him to the casino one evening. The two men drove in Bjoern's car, talking statistics.

"I don't gamble," Bjoern explained to Ted. "I only play blackjack and I use a system that by memorizing the cards permits me to change the odds from 10% against me to 3% in my favor. It is hard work and I never win big. On the average I make in an hour about the same as with my salary at CERN. What I enjoy here is knowing that the croupiers think of me as an addicted gambler who will eventually lose like everyone else, while I know I am winning."

Bjoern liked to impress Ted with his knowledge of mathematics and statistics. "Did you know that there are more people alive today than have ever died?" he asked rhetorically.

"It seems hard to believe," pondered Ted.

"It is a mathematical consequence of the fact that the Earth's population has been increasing exponentially for the better part of its growth process. The statement actually holds true only if you begin counting around the time of the construction of the Egyptian pyramids, i.e., 5,000 years ago. But earlier world population estimates are too small and too speculative anyway."

Ted had a statistical curiosity of his own.

"One day I proved to myself that it is abnormal to be normal. To hide in a crowd, it suffices to look like everyone else, that is, be of average height, have average weight, and generally adhere to the banality of commonplace in all respects. But how common is it to come across a truly commonplace person? In psychology normal is defined as being close to the average. Concerning IQ, for example, 68 percent of the people are defined as *normal*, meaning their IQ lies within one standard deviation from the average. The same can be said for people's height. However, only 47 percent of the people—that is just about one in two—are within one standard deviation in both IQ and height at the same time. Suppose now that we are trying to find someone who is normal also in weight, in wealth, looks, in health, in sports performances, and so on. It turns out that one person in three is normal in three different aspects simultaneously, and one in five is normal in four aspects. But the probability goes down rapidly the more aspects you want to be normal in. To be normal in ten different variables the chance is 2 percent. There is only one chance in a million that someone is within one standard deviation in 36 different variables simultaneously, which makes him or her a real freak worthy of a place in a circus. And there should be no one on earth today who is normal in 55 or more different aspects."

Bjoern was not surprised by Ted's argument but had never made the calculations himself. "In this light," he concluded, "We must revise the definition of a normal person. A person should be 'normal' when he or she is close to the average in some aspects but far from the average in most other aspects."

"In fact," continued Ted, "I have calculated the weighted average as 3.2 variables and to lie within one standard deviation from this average it suffices to be 'normal' in two to four different aspects, but not more or less. For example, if you are average only in intelligence and height, you are 'normal.' But if in addition you have average will power, and your income is also average, then you are not quite normal."

They laughed by their conclusion that being different rendered people normal.

Once inside the casino, Ted found the large games area was filled with smoke, lit dimly with only incandescent yellowish lights, and insulated from the outside world by heavy dark red velvet curtains on all huge windows. Some men were dressed in smoking jackets and many women wore long dresses revealing their shoulders.

Bjoern walked slowly toward the blackjack tables.

"I will take a seat and play continuously," he told Ted. "If you see me raise my bet, it means that the count has turned favorable so you can add your bets on top my hand, if you want; it is allowed. But I will pretend I do not know you."

Bjoern was very careful not to raise suspicions with the croupiers concerning his counting of the cards. "If they suspect that I am counting the cards, they will ban me from the casino," he had explained to Ted.

Two hours later Bjoern began showing signs of fatigue. "Concentrating becomes more taxing if you have to conceal it," he confided to Ted later. Ted bet on Bjoern's hand on several occasions, but at the end they both finished with only modest gains. They went to the bar for drinks.

Ted took the opportunity to talk to Bjoern about esoteric work and the Gurdjief group, but found him unreceptive. Bjoern was a stanch Catholic and anything spiritual had to be related to the Bible. Ted seemed annoyed.

"How do you reconcile gambling and living with two women simultaneously with being religious?" asked Ted provokingly.

"There is nothing in the Bible against gambling," he replied. "And when I was living with two women I was not being unfaithful to either of them. You should not take the Ten Commandments literally. For example, there are situations where lying is condoned as it promotes the common good. Do you know the solution of the ballroom-dilemma problem?" he asked enigmatically.

"What is the ballroom-dilemma problem?" asked Ted.

"In older times during high-society balls the ladies had a dance card on which they noted down the requests for dances. Typically the men would come in the ballroom one by one and go directly to the most beautiful lady asking her for a dance. She could accept the request and note it down in her card, or she could reject him for some such reason as too many commitments. A man also had the option of first addressing the second most beautiful lady in order to make the most beautiful one jealous and perhaps make her more likely to accept him later. The question is who should connive and when, or what merit could there be in lying?" Bjoern was obviously leading to the solution he knew of an otherwise complicated mathematical optimization problem.

"The problem has been rigorously solved," he continued. "But the solution does not aim to satisfy one man, or one woman, or even one couple. The problem has been solved so as to maximize the collective well-being. That is to make the whole group of men and women as satisfied as possible."

"I bet the women should lie," said Ted with a grin.

"Yes, but it has noting to do with the gender," insisted Bjoern. "The person who asks needs to be honest, and the person who is in demand should connive, if the global happiness is to be maximized.

"Even today when girls play hard to get, they are not simply being picky. Unwittingly they are making their contribution to maximizing the well-being among all youngsters in the neighborhood," concluded Bjoern triumphantly.

Years later Ted had the opportunity to go into depth with this problem. Together with a mathematician subordinate they incorporated the mathematics into policies used by headhunters and placement agencies. It turns out that when you apply for a job and there are many candidates for few posts, it is recommended that you be honest. On the contrary, if there is high demand for your skills and you have little competition, you (but also the community) will benefit if you sacrifice frankness while making your choice.

In retrospect Ted thought the mathematical solution sounded very reasonable. People seem to have subconscious knowledge of it because they generally behave like this. Here again was a case of a truth proven by mathematics but known to ordinary laypeople intuitively.

16 – The Power of Attention

Ted often became bored as he sat in the CERN auditorium listening to physicists presenting their results. On such occasions he felt taken advantage of because he "wasted" his attention. The same thing happened at uninteresting concerts and poor artistic performances. He thought he was paying twice at concerts and plays, once at the box office and a second time during the performance with his attention. He considered his attention as a precious resource of which he had a limited amount and therefore should use it wisely. He would not forget the gratitude of his best friend in high school when he wanted to recount to Ted an intense emotional experience in detail as if to relive it. Ted listened to him attentively for a long time, at the end of which his friend was filled with warm emotion for Ted. He said that the attention Ted had given him was tantamount to love.

On the other side, whoever gives a talk or performs on stage has the sought-after opportunity to draw attention in massive quantities. Physicists, just like actors and performers of musical instruments, cherished the attention received via the limelight. At first Ted thought it was an ego trip and for physicists he considered it another case of "humanities of science." But little by little he came to recognize that attention is indeed a valuable commodity that possesses exceptional powers.

Attention has nourishing qualities and can achieve spectacular results. You put your attention on fine glassware and it will not break. You give your attention to a plant and it grows healthier. You give your attention to your children and they do better at school; to your spouse and it makes your marriage last longer.

Drivers never become carsick, only passengers do. Whenever Ted drove his children on winding roads he instructed them to pretend that they were the drivers. This is because the attention and the concentration focused on driving prepare one's body for the movements that follow so it does not swing uncontrollably as the car moves. A similar thing happens in hiking or shopping. The person tagging along easily gets tired; not so the leader or the one who is shopping. Ted used to put the most tired child in charge of finding the fastest way home.

The power of attention was used extensively in the Gurdjieff group. One of the early exercises Michel de Salzmann kept coming back to was to use one's attention in order to "animate one's body," as he put it. If you lie still as if you were going to sleep and concentrate your attention on your left arm, you will eventually feel warmth permeating this arm. If you put your attention on your lips, they will feel warm; if you persist, they will feel tingling and almost numb. The complete exercise aimed at getting one's entire body in this "animated" state by circulating one's attention over all the parts of the body.

When Ted did these exercises he thought that blood rushed to the part under attention and gave rise to these sensations. But he could not understand why attention had such an effect on one's blood circulation. Could it be that attention relaxed the muscles opening up blood vessels to receive more blood? One way or another, it was clear that a non-physical stimulus directed and controlled exclusively by one's mind could achieve concrete physical results even on inaccessible inner organs of the body. Ted reclassified some of his "weirdoes" related to mind-over-body dominance into drawers with fewer question marks, for example, the myopic person who under hypnosis did not need his glasses to read a distant clock.

But getting hold of someone's attention can also have devious effects on the person. Hypnotists use some means (a pendulum, a shiny object, a finger, and their voice) for grabbing their subject's attention, which they hold on to and manipulate afterward. It is trivial to hypnotize a chicken. Put the chicken on a flat concrete pavement and with a chalk begin drawing a straight line on the pavement making sure that the chicken sees you. The chicken will fix the line and will follow it obediently as long as it unfolds in front of it.

The snow pattern on an empty TV channel or other electronic-noise pattern can also produce a hypnotizing effect on the person who directs his or her attention on it. The screens of television or computer monitors are in any way sinks of attention. After all, the Internet's most precious commodity is people's attention.

Marshall McLuhan had made a big point out the different kinds of attention involved in watching television and going to the cinema. There were no high-definition TV screens at the time and the TV picture was not as clear as on movie-theater screens. TV watchers "participated" in the viewing process by supplying the missing-resolution detail themselves, something that would affect their long-term evolution and development.

But there is another difference between television and movie theaters that affects the watcher's involvement and enjoyment. It has to do with the fact that in cinemas it is dark and quiet, and there is nothing else to do but watch the movie; watching the movie demands and obtains one's exclusive attention. Not so in the living room. TV watching can be and is interrupted for many reasons besides commercials. The eyes easily drift away from the TV screen. As a consequence the same film can result in a completely different experience watched on TV than in a movie theater. The quality of attention involved changes the experience. In fact it can even change what one sees. Proof lies with Escher's ribbon cube, where some of the cups can be seen as projecting outward or receding inward depending on how you look at them.

With the right kind of attention old familiar surroundings can appear as new. Because of their nutritive properties Gurdjieff considered new impressions as a kind of food for humans and as such it should be a limited resource, but there is no limit to the amount of impressions available. Impressions become a limited resource because they are transmitted through one's attention, which is limited. Without attention, impressions are useless, and without impressions life ceases.

YOU CAN CHOOSE WHAT YOU SEE, CUPS PROJECTING OUTWARD OR RECEDING IN

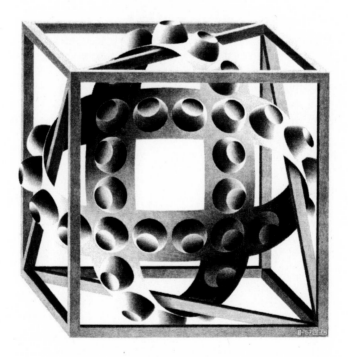

Cube with Magic Ribbons, by M.C. Escher (*lithograph, 1957*).
Image Source: attanatta / flickr.com. CC BY 2.0
(https://creativecommons.org/licenses/by/2.0/)

Ultimately impressions may not even be necessary; a high-grade attention alone can provide essential nutriments. Highly developed individuals might "nourish" themselves on stimuli coming only from within. People such as yogis and martyrs can survive and even be serene wherever and under whatever conditions they may happen to be.

Cheap thrills can be thought of as "loud" impressions that one receives with minimal (low-grade) attention. But the effect of a strong stimulus on low-grade attention is smaller than a weaker stimulus on high-grade attention. Watching a Spielberg film on TV leaves a lesser mark behind than discovering Bach's genius by listening to his music intently. Furthermore, because attention is dirigible you can see what you want instead of what others may be pushing on you. On a rocky beach of northern Scotland Ted once found a large flat pebble smoothed by the waves to reveal stratified minerals lines on the stone's structure. He was in a euphoric state and looking at the unspecified patterns on the stone he was able to "read" the better part of a love letter. He was bewildered. He couldn't believe how clear the writing seemed to him; he had trouble with only a few words where not enough letters were visible. He took the stone with him and went back to it many times in the following years but was never able to reread that letter. The setting and other conditions on that remote Scottish beach had provided him with an attention of unique quality.

Euphoria occurs when attention is heightened to the level of enabling someone "see" things with minimal stimulus.

Hallucination occurs when the attention sensitivity is so high that no external stimulus is necessary; it gets triggered by noise only from within.

It has been argued in mystic literature that attention can be of benefit even to the dead. The astral body, which is supposed to take over when one's physical body disintegrates, is nourished by the attention of living human beings. That is why gurus and religion founders cultivate a cult of worship. After they die the continuous beaming of attention by the faithful will benefit their astral bodies. But Ted had buried the concept of an astral body deep in his weirdoes filing cabinet into a drawer with many question marks on it; not so with attention.

Attention has also been used as a tool to approach the more nebulous aspect of our consciousness. One of Gurdjieff's fundamental consciousness-raising exercises was directing your attention to yourself by "remembering yourself" *at all times*. This exercise proved a challenge with which Ted struggled. But earlier he had been through a similar type of exercise with Aris and Mihali during a Mendios get-together. They had directed their attention at a corner of the ceiling of the room in which their meeting was taking place and tried to perceive themselves from that perspective for a prolonged period of time. Concentrating one's attention outside one's body makes one conscious of (and thus fill) a greater space.

Mihali had gone further. He had exercised on maintaining conscious and continuous knowledge of where the sun and the moon happen to be throughout the day.

Ted's date had been taken aback by his reaction one summer evening while they were dining romantically in a restaurant on the Lake of Geneva. Looking at the bright full moon in the sky she had asked him what it made him think of. Without hesitation he replied, "That the sun is right behind our backs."

She had expected something more romantic.

17 – Love Science

The next summer camp of the Gurdjieff group was scheduled to take place exceptionally not in Chandolin but in Amsterdam. Ted decided to go by train, which takes about sixteen hours from Geneva. He began his trip in the afternoon with a sleeping-car stub on his ticket. He enjoyed the continental European trains with the closed compartments and the fancy restaurant cars. The seats in his compartment could be transformed to beds and this way he could sleep during a large part of the long trip.

When you first get on a train for a trip like this there is always a moment of concern. With whom are you going to share the small compartment, probably for many hours? The person who sat across Ted was a distinguished-looking man in his fifties. His carefully trimmed beard and somewhat old-fashioned clothes made Ted think of an impoverished aristocrat of older times. He soon found out that his guess was not far from the truth.

Gilbert Coomans de Gryse was Belgian, indeed of aristocratic descent, now working as a civil servant at the national statistics office in Brussels. He would share the compartment with Ted for about five hours before changing to another train. The two men began chatting and wasn't long before de Gryse revealed his eccentric prejudiced character. Ted remained generally polite but was not entirely happy with his interlocutor's antifeminism.

"Women today receive what they deserve," announced de Gryse in a rhetorical way. "In older times a woman's main concern was to get married. There was so much pressure on women to get married that they made outrageous concessions toward that goal. But that led to bad marriages because once married they could not maintain the level of concessions. As a consequence marriages saw husbands neglect and mistreat their wives, who were no longer being as nice to them as before.

"Women asked for and got more rights so that divorces became possible and easy, and they began divorcing their husbands and claiming alimonies for themselves and to raise the children. Later, with more equality, women began working like men in the marketplace, which deprived divorced women from the personal alimony. Also the pressure for a woman to get married eased off as feminism rendered the concept of single mothers acceptable. But this deprived women from the alimony to raise the children. So today's women have to work like men, have children on their own, and raise them on their own. Other than that, feminism has triumphed, Madame la Marquise."

Ted and de Gryse were speaking in French and de Gryse's last remark alluded to the popular French folkloric song about a Marquise who returns to her châteaux following a long absence. Her valet announces to her all sorts of catastrophes, one by one, adding each time, "other than that everything is fine, Madame la Marquise."

"I don't think feminism argues against marriage," Ted tried to inject some political correctness in the discussion.

"It doesn't need to, my young man, couples break up these days for any odd reason: because they fall out of love, because they were not *really* in love to begin with, because they are too different and disagree on too many issues, or because they are too similar and therefore get bored with each other." De Gryse let his cynical self come out.

"Actually you are right," Ted took the opportunity to redress the level of the conversation. "Demanding careers are all too frequently associated with a couple's separation. Nobel physicist George Charpack confessed publicly that his research activities dictated the evolution of his love affairs. His is a typical case among dedicated researchers. But at the same time, couple therapists criticize symbiotic relationships, that is, relationships in which there is 'pathological' interdependence, stemming from financial, emotional, or other insecurity. An individual's autonomy and independence are consi-dered as necessary ingredients for forging a healthy couple in today's envi-ronment. Yet, it is this very independence that offers the degree of freedom permitting separation. A simple golden-mean type of rule, meaning some-thing like having enough independence but not too much, appears easy to say but difficult to apply."

"This is not a science, young man." De Gryse had been impressed when he found out that Ted was a particle physicist. He was at least ten years older than Ted and he insisted calling him "young man," as if to counterbalance Ted's authority stemming from the fact that he was a scientist.

"And yet, I can see some science in it," objected Ted. "Think of Newton's law of the falling apple. He established this law by watching apples fall, but falling rocks also obey the same law. Even when we throw a stone, it may move horizontally, but at the same time it follows vertically an accelerated fall like that of a falling apple. That is why it traces a curved trajectory to hit the ground a certain distance away. If we throw the stone horizontally with more force, it will travel farther, tracing a flatter trajectory, before hitting the ground further, but always falling like an apple. Imagine now throwing the stone with an immense force so that the curvature of its trajectory is as flat as the surface of the earth. The stone will keep falling downward, but it will be tracing a trajectory parallel to the surface of the earth, therefore never touching ground. This stone will become a satellite of the earth. You can think of the moon as continuously falling toward the earth like an apple, but because of its large horizontal speed it keeps going around never reaching the ground. Similarly the Earth can be thought of as falling toward the sun, and in general any two astral bodies revolving around each other are 'falling' toward each other according to Newton's law.

"An interesting side effect of free fall is that the pulling force is no longer felt. You feel the gravitational force on your feet pressing the ground when you are standing still. If you embark on a free fall, you no longer feel the force of gravity. For the people in an elevator that is falling freely because its cable snapped, the earth does not exist (at least not until the elevator hits the ground!) They are weightless while freefalling.

"Applying Newton's falling-apple law to social situations may be an overly simple-minded exercise, but there is something to learn. We can think of two people who are attracted to each other as 'falling' towards each other. (Appropriately we speak of them as *falling* in love.) If there are no obstacles or other mechanisms that interfere with a 'free fall,' the two people will get closer together but at the same time will stop feeling the attractive force. Much poetry and prose have been devoted to the ephemeral aspect of falling in love. On the other hand, obstacles blocking the two people from getting together will maintain the feeling of the attractive force. Romeo and Juliet's attraction toward each other did not weaken with time; on the contrary the obstacles in their relationship intensified it until it drove them to killing themselves. But the attractive force can be responsible for more honorable achievements, provided that free-falling motion does not water it down. It has been argued that Dante would have not written the *Divine Comedy* had he consummated his relationship with Beatrice."

Ted was getting excited and de Gryse also seemed amused. "Have you got more of this?" de Gryse asked with toned-down sarcasm.

"I do," replied Ted smiling. "A more sophisticated physics law for the attraction between two objects is Coulomb's law governing electric charges. It is like Newton's law of falling apples but in addition we now have the charge differentiation that produces attraction between opposites but repulsion between the like. Electromagnetic forces constitute the best sociobiological equivalent for sexual attraction. No wonder romantic writers use jargon such as magnetic attraction, current flowing, electricity, and charged atmosphere in love stories.

"Coulomb's law can yield some insights concerning social phenomena. At first sight it seems to say that complementarity results in attraction while similarity in repulsion. This in general makes sense, for example, sexual attraction shows up best between oppositely polarized sexuality. Archetypically, the more he-man he is and the more she-woman she is, the greater the attraction.

"However, there are elaborations on the stability of a couple coming from quantum mechanics. The Pauli exclusion principle states that sameness gives rise to a repelling force. The electric charges (the sexes) do not necessarily have to be opposite, but some differentiation is essential for stable coexistence. Physicists talk of a *Pauli force* or *pressure* that pushes identical particles apart. The ancient Greek philosopher Hesiod first documented how sameness results in competition in the 8th century BCE:

> Potter is potter's enemy, and craftsman is craftsman rival; tramp
> is jealous of tramp, and singer of singer.

"If Pauli's exclusion principle has any social validity, same-sex couples should be unstable unless differentiated. Because the higher the resemblance, the more the partners are likely to fly apart (possibly via competition). Heterosexual couples are sufficiently differentiated by sex. But homosexual couples may have a problem if the two members are also identical in other respects. The most unstable couple would result from identical homosexuals. Barring forceful confinement—as coming from being stranded on a desert island—identical homosexual couples are doomed by the Pauli exclusion principle (technical confinement can help to some extent such as a business partnership or other common project.) Two protons will not stay together all by themselves. And yet, it is possible to render a two-protons state stable by adding a neutron, which is a particle with no charge and therefore no electric interaction with the two protons. The proof lies with Helium-3, which is a stable isotope of helium that has in its nucleus two protons and one neutron.

"The existence and stability of Helium-3 tells us that two identical homosexuals can form a stable couple in the presence of a third *neutral* party. There should be no interaction of sexual nature between the homosexuals and the third party. Ideally, this neutral entity can be a child, a parent or older person, or even a pet of importance.

"On the other side, oppositeness and complementarity produce couples that can in principle achieve high stability. This is the case of stereotype heterosexuals, or amply differentiated homosexuals. The physics analogue is a proton and an electron forming an atom. The particles are very different and oppositely charged. The more tightly bound the atom—the closer the opposite charges are physically—the lower the energy content of the system and consequently the more stable the atom will be. There is a danger, however. Binding an electron too tightly to a proton may create a neutron particle, thus destroying the atom. In contrast to an atom, in a neutron the positive and the negative charge components have lost their individual identity. In stable atoms there is some kinetic energy—or diffused existence —for the electron so that it does not collapse on the proton. The social understanding is that at least one member of the couple must have some activity of importance outside the couple, a career, a peer group, or own friends and interests.

"In the case of *extreme* complementarity, we are facing a situation that corresponds to yet another type of pairing in physics: matter and antimatter. There is attraction between a proton and an antiproton, but when they fall toward each other, they collapse and annihilate into large amounts of energy in the form of an explosion. The social equivalent is obvious: the ultimate he-man meets the ultimate she-woman. They are immediately attracted to each other, fall in love, and if left alone, collapse in a doomed short-lived relationship that will go up in smoke either via some degenerate interdependency or some form of terminal sex. Extreme complementarity can be detrimental. The way to maintain stability in such a couple can be found in the proton-antiproton atom. This is a stable configuration of a proton and an antiproton circling rapidly around each other. Both particles have a sufficient amount of kinetic energy to prevent them from collapsing on each other. What it means for a human couple is that they should both have strong independent reasons for existence, e.g., demanding careers. But once again, there is a limit. Too much kinetic energy will make the proton fly away from the antiproton like a satellite that was given a velocity greater than the escape velocity. This could be a career that suddenly becomes so successful that it transforms one partner into a star. Evidently two stars (movie stars are a good example) would have too much kinetic energy to stay together for any reasonable amount of time. Look at Hollywood; it is extremely rare to find two movie stars that make a couple stable in the long run.

"To summarize, opposites attract each other and as they get closer together their overall energy becomes lower thus producing a more stable state, but if the opposites are extreme, or if they get too close, the situation can turn explosive. What looks like a nice recipe for togetherness can easily deteriorate into disaster. Society and religion have tried hard to introduce additional binding mechanisms for couples (marriage legalities, divorce stigma, etc.) But as these institutions weaken, the survival of couples depends on the extent to which they benefit from the right admixture of the appropriate stabilizing behaviors. To add stability the rule of thumb is: strengthen similar couples with common goals and a third-party presence; and balance complimentary couples with personal interests and independent careers. The career demands will keep them running and despite the irresistible attraction will prevent a fatal collapse.

"A possible contradiction may seem the facility with which identical twins get along together. But identical twins are not in a sexual relationship. They enjoy binding forces of a different (genetic) origin. There is an analogue in nuclear physics. Two neutrons are pulled together by the nuclear force, which is different (weaker) from the electromagnetic force. As a consequence, the dineutron is *almost* stable. Analogously, identical twins do not generally live their entire lives together."

Ted was impressed by his own performance. He had thought of these things before. But his lengthy exposition brought him a euphoria that acted as a catalyst. He improvised and enhanced the ideas as he presented them. De Gryse also listened attentively.

"Are you writing a book about all this?" asked de Gryse.

"No, but I may do so some day," answered Ted, even though the idea had not crossed his mind.

18 – The Opposite of a Great Truth Is another Great Truth

The old farm close to Amsterdam where the Gurdjief camp took place no longer resembled a real farm. The various buildings had been transformed into facilities such as dormitories, meeting rooms, woodshops, metal shops, storerooms, and gym floors. There was even a large old-fashioned outdoor oven posing as a monument in one of the yards. Ted soon found out that it was functional and that the daily bread needed for the meals and collations was baked in it.

There were many new faces. The first evening Ted met Peter Brook, the well-known English play and movie director. Peter was treated as one of the elders in the Gurdjief hierarchy and at dinnertime sat at the head table next to Michel, José, and a handful of others facing the 120-strong audience. Ted knew of Peter from cinema. As a college student during the sixties Ted had been impressed by the music in a French movie called *Moderato Cantabile* that Peter Brook had directed. It was a sweet piano melody that sounded simple enough for Ted to play on the piano. But he did not know how to find the music score. In a desperate attempt one day he walked into a music store downtown Manhattan and asked the storekeeper, "I wonder whether you can help me. I am looking for the sheet music of a piano melody that goes like this: ta, ta, taaa, ta, ta, ta, taaaaa…" The storekeeper smiled but recognized it. It was from Albinoni's Sonatina No 4.

Meeting Peter now in the flesh and in fact as one of the Gurdjieff elders was an emotional event for Ted. Moreover, they could speak in English, which was something that made Ted more comfortable. Michel introduced Ted to Peter as "the physicist." Peter did not blink an eye but it wasn't long before Ted and Peter found themselves drifting away from the others during collation and other free time to talk about all kinds of issues. Ted as usual had questions, lots of them, and mostly intellectual ones. Peter kept a down-to-earth attitude.

"What would you do if you found out the exact location of a monastery in the Himalayas where they have *real* knowledge and can answer your questions about God and the purpose of life?" Peter asked Ted provocatively.

"I'd rush there," replied Ted, wanting to play up to Peter.

"No," Peter said disapprovingly, "you should stay right here and keep doing the Work as Michel directs it. The answers cannot be given to you by some expert. You have to find them out by yourself through hard work, the kind of work we are doing here. They would tell you the same thing there, and why go through all the trouble to seek such a remote place if you have an alternative right here?"

On another occasion Ted complained to Peter about the lack of ideas and intellectual stimulation during the daily activities at the camp and in the Gurdjieff groups in general.

"What ideas are you talking about?" Peter seemed surprised at Ted's question.

"Ideas like those in Ouspensky's book *In Search of the Miraculous*," answered Ted, reminiscing nostalgically his days in New York City with Mihali and Aris.

"Those ideas have only one reason for existence," continued Peter, "to lure people like you into the Work. Once you have been caught, like fish in the nets, you must forget about the ideas and begin Working. You have done enough thinking in your life. Now is time for you to stop thinking and begin being."

Peter spoke calmly but with conviction. His last remark touched Ted profoundly. It came as a confirmation. Ted was aware that he had paid excessive attention to his intellect in his life. In a Mendios meeting with Aris and Mihali they had expanded on Gurdjieff's idea that human beings consist of three "bodies" a physical one, an intellectual one, and an emotional one, and how the harmoniously developed person—in contrast to the typical westerner—should have all three equally developed. Associating intellect with the head, emotions with the chest (heart), and physicalness with the lower body, Ted and his friends had drawn bean-shaped caricatures for themselves. All three caricatures turned out to be by far top-heavy.

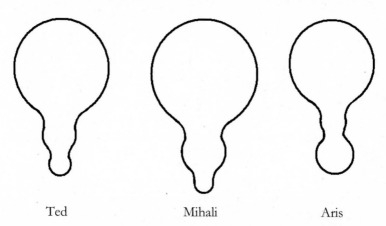

| Ted | Mihali | Aris |

For the rest of the week Ted found himself progressively abstaining from spontaneous participation in group activities. Peter's words kept buzzing in his mind. They often mingled with images of the new car he had recently ordered and was now waiting for him to pick up at the BMW factory in Munich.

Soon after his return to Geneva he decided to drive his brand new BMW to Paris to visit Marie-Louise, the violinist he had met in the first camp at Chandolin.

Marie-Louise played the violin in the Orchestra of Paris, and when Ted visited her in Paris he attended one of her concerts. He sat alone among strangers in the audience while she played on stage with the other musicians. He enjoyed it very much. Not only did he like classical music but he also used the time as opportunity to think, meditate, and make scientific observations.

"I got the impression that the conductor was following rather than leading the orchestra," he told her after a concert. "Are you sure that he is indispensable?"

She agreed. "Yes," she admitted," if you took this conductor away, you'd hardly notice any change in the sound of the music. This is the case with many conductors. But some conductors are effective and their presence makes a tremendous difference."

Ted was bemused. "So here is a case where opposites meet," he said. "In a concert the most central character, the orchestra conductor, can be any-where from having zero effect to having a tremendous effect."

They decided to go for dinner and as they were not far from the Eiffel Tower Marie-Louise suggested that they go there. Ted did not know Paris very well but had visited the Eiffel tower and did not like being a tourist. Marie-Louise reassured him, "Don't worry, it is not going to be corny."

They took the elevator to the first floor of the Eiffel Tower. Ted had not been aware of a rather nice restaurant there. It was spacious, with view of Paris in all directions, and strangely it was not full of tourists. Once they had settled at a table and ordered, Ted returned to his idea of how opposites meet. He gave Marie-Louise a lecture.

"The great Danish physicist Niels Bohr said that there are two kinds of truth: simple truth and fundamental truth. The opposite of a simple truth is a lie; the opposite of a fundamental truth is another fundamental truth. And that is how one can distinguish simple from fundamental truth.

"Whimsical as it may sound the statement has merit. To say that I went shopping today excludes the possibility that I did not, given that I am truth-ful. If a statement is true, its opposite is false. But to say that love is all you need is equally true as its opposite, namely that you cannot live on love alone.

Having gone shopping today is a simple truth. Love is all you need and its opposite are fundamental truths.

"If you can turn it around, it must be a fundamental truth. Proverbs and popular sayings typically express fundamental truths and therefore can be turned around. For example, no news is good news. At the same time, its opposite, namely good news is no news, is also true, if you ask a journalist.

"The more fundamental the truth, the more its opposite will also be true. People generally agree that chewing your food well is good for you. However, the argument has been made that your stomach may atrophy if your food is constantly over-chewed and the stomach does no work. That would be bad for you.

"It is generally accepted that hard work can break one's back. Porters carrying large loads are known to often suffer from disc hernias. But at the same time it is also known that what does not kill you makes you stronger. The strongest backs can be found among people who indulge in lifting overly heavy weights.

"This is why there exist many contradicting proverbs: 'Opportunity seldom strikes twice' but 'all good things come in threes.' 'He who is in a hurry always arrives late' but 'he who hesitates is lost.'

"Oscar Wilde may have said it in jest: 'There are two tragedies in life; one is not to have what you want; the other one is to have it.' He was referring to his tumultuous and passionate relationship with a young man, which resulted in tragedy both when they were together and when they were apart. Oscar Wild was telling the truth as well as echoing Bohr's idea. Finding truth in the opposite of truth is invoked twice here, once by the phrase's content and once by the fact that it is a joke that turns out to express a serious truth, as is often the case with jokes.

"Ancient Greek wisdom argued that there is nothing bad without something good in it. But Gurdjieff extended that to: everything bad has an *equal* amount of good in it and vice-versa. The greater the misfortune, the greater the benefit hidden behind it. The more beneficial something seems to be, the greater the calamity that will eventually stem from it. The challenge is to identify the hidden opposite.

"I tried to make this point once in a social gathering and a vocal objection was raised by a Jewish scientist, 'What was the great benefit associated with someone like Hitler?' she asked.

"At first glance Hitler seems to be a source only of calamities, especially for the Jewish people. In retrospect, however, Hitler can be held responsible for the Jewish Exodus and the creation of Israel. A people of Diaspora for 2000 years finally got a long desired state. Israel would have most probably not have been created had it not been for Hitler's persecution of the Jews.

Granted, Hitler provoked a great catastrophe but also triggered a great benefit to the Jewish people."

Marie-Louise appreciated what Ted had to say, but she could only take so much of his lecturing at a time. She made a remark about his food getting cold. He understood and even though he had more to say he shut up. He had been carried away again with one of his "discoveries." He could not help it. Every time he unearthed some unconventional truth or other unfamiliar piece of knowledge he felt he had made a discovery and wanted to share it with those close to him.

19 – Opposites Meet

It took anywhere between five and six hours to drive back from Paris to Geneva even with a BMW. But Ted enjoyed driving. He had equipped his car with a good radio/cassette player and many cassettes with his favorite classical music. The conditions while driving—the steadily changing scenery, the murmur of the motor, and the music—were conducive to contemplation. He had invested in a high-quality tiny micro-cassette recorder. He used it as a notepad to keep track of ideas at euphoric moments. On this trip he kept thinking about the topic he had lectured Marie-Louise on about the meeting of opposites. There were so many things he did not tell her. He turned on the micro-cassette recorder, let it run recording on the passenger seat, and tried to forget it while putting his thoughts in order.

Thinking big, Ted asked himself what could be the opposite of the Big Bang. Well, it should be something like the Ultimate Collapse, a possible end of our universe, which has not has yet been ruled out. Physicists talk about the possibility of an eventual coagulation of all matter (and energy)—after they become uniformly distributed throughout the universe a long time from now—into one densely concentrated state. From then onward another Big Bang is likely. There are theories that our universe has been preceded (and will be followed) by other universes. The end of the Ultimate Collapse of one universe is the beginning of the following universe's Big Bang. This is one grandiose case of extremes that meet.

Mathematics has its own way to deal with extremes that meet. Zero and infinity have many properties in common and there is a special relationship between them. The existence of one implies the existence of the other. You can turn any number into infinity by dividing it by zero. And vice-versa, you can turn *any* number into zero by dividing it by infinity. Consequently, every number is mathematically equal to zero times infinity, which invokes a philosophical image with eastern-religion overtones, namely that you can find nothing and the whole world in everything.

To find nothingness in oneself one needs not invoke religion or philosophy; physics alone suffices. One's body is made of cells that are made of atoms, each of which is made of a nucleus and electrons around it. But most of the material in an atom is concentrated in the nucleus. If this nucleus were the size of a tennis ball, the size of atom (defined by the position of the furthest electron) would be six miles in diameter. Consequently, the empty space in an atom corresponds to 99.9999999999999 percent of its volume. We are, just as everything else around us is, full of empty space. The reason we do not fall through the floor—both the floor and we are mostly empty space—is that nuclei are positively charged, and there are strong repelling electric forces between them that have no trouble manifesting themselves across empty space. In fact, electromagnetic forces are responsible for almost all shapes and phenomena observed on the earth. At the same time, the whole body can be found in any of its parts. Each human cell carries all the information needed (DNA) to reconstruct (clone) the whole person.

Mathematics has also linked intelligence to fallibility. Several theorems say that if a machine is expected to be infallible, it cannot also be intelligent. Unfortunately, these theorems say nothing about how much intelligence can be expected from machines that do make mistakes. But one way or another, there will be no threatening competition to human intelligence from powerful computers as long as the latter are supposed to make no mistakes.

Opposites meet in yet another way in the human body. What we eat can be a poison or a medicine at the same time. Contaminating an organism with viruses and poisons in small quantities constitutes the essence of many curative techniques, such as vaccines and homeopathy. It has been argued that even small doses of radiation and toxins can be beneficial and increase longevity. At the same time, nutriments as benign as sugar, pasta, and proteins can become poisonous when ingested in excessive quantities.

But opposites also meet in social life. It may not be surprising to witness such contradictory behavior as laughing when something becomes too serious, or feeling relief when a big new problem is added on top of an already problematic situation.

A car damaged in an accident can be repaired to perfection. A mechanic working in an auto body shop can justifiably boast, "No matter how badly damaged a car comes into our shop, when it goes out it looks as if it just came out of the factory." It did not take long for Ted to discover the exact opposite. No matter how good the repair job, one can always tell that a car has been through an accident by looking carefully and at the right angle.

A negative statement can imply something positive. For example, when you ask what the weather is like outside and someone answers, "It is not raining," you understand rather that it is about to rain. Someone who tries to convince others that he or she is not a racist—the telltale phrase is "but some of my best friends are …"—reveals a racist disposition. In a sort of reversed causality, the existence of an answer makes the question legitimate. It is the answer that raises the question. For this reason sensitive people refuse to answer questions on their honesty, integrity, and other things they hold dear.

Another type of reversal causes someone to be "born" right after he or she dies, through publication of the death announcement. When large-circulation newspapers carry the announcement of a literary person's death, the number of people who just find out that this person existed can well surpass the number of people that knew of this person when he or she was alive. In that sense death can make one more "alive" than life. Disturbed personalities may exploit this in order to go down in history.

It can also happen that opposites meet by convention. Two people may see colors differently. One sees as red what the other sees as green. Neither is colorblind and there is no confusion because they have learned to call the colors with their conventionally correct names, independently of how they *really* perceive them. One may see green as red, but since he has learned to call it green, no one can know that he really sees it as red, not even the person himself. The statement can be generalized to claim that there are no two persons in the world that see colors in the same way. And raising the level of abstraction, one could generalize colors to shapes and forms, meaning that no two persons in the world see something in the same way. This thinking constitutes more than an intellectual exercise. One can argue that these are the origins of art.

Opposites also meet on a much larger social scale. War manifests itself both at the peaks and at the valleys of the long economic wave—otherwise known as the Kondratief cycle—as if both hardship and prosperity were valid reasons for it. World War II took place during depression years, the Viet Nam war during economic-boom years. The needs of these periods were different. The former may have been survival-driven, whereas the latter greed/expansion-driven. The student revolution of 1968 seems unfounded compared to revolutions of the repressed, such as the French and the Russian.

Children of criminal parents, with no family support, money, or education, may become delinquent, *like* their parents. But so may the children of well-to-do families who provide them with everything, *by reaction*. Children who are given everything may not be motivated to work toward something.

The simple existence of something becomes the reason for the existence of the opposite. Affluence often causes complacency, reduced productivity, and even decadence, whereas penury triggers innovation and creativity. Something provocative and groundbreaking can be easily seen as obvious and commonsense-like. And vice versa, rock-bottom common sense can be mind-blowing. Beauty can only exist in contrast to ugliness, and good in contrast to bad. You cannot discern what a picture depicts if the foreground and the background are the same. Difference—and par excellence oppositeness—is key to existence itself.

IS IT THE PRESENCE OF THE DEVIL THAT CAUSES THE ANGEL TO APPEAR OF VICE VERSA?

'Circle Limit IV' (Angels and Devils), by M.C. Escher (*woodcut 1960*)
Image Source: attanatta / flickr.com. CC BY 2.0
(https://creativecommons.org/licenses/by/2.0/)

Innovation is by definition linked to the unexpected. More often than not a headline-making new theory is simply the reverse of the accepted theory. There are many examples. In anthropology Darwinism based on competition, has been under attack from the post-Darwinists now arguing for cooperation and co-evolution. In sociology the perennial discussion of nurture-versus-nature argument has seen swings from one side to the other. For decades academic opinion held that early-age development was solely responsible for a person's later behavior. But then genetic explanations became the fashion.

Granted there has been new knowledge, more research results and under-standing on how molecules, micro-organisms, and genes work. But part of the reason for the swing to the opposite direction is that a change was over-due. A change of mentality offers fertile ground for constructive new work.

Ted concocted a "recipe" for innovation: find a well-established trend that shows signs of fatigue (there is a critical evaluation in this with respect to timing). Reverse the fundamental premises and launch it as the backbone of a new direction for research.

Some people do this on an everyday basis. Individuals who are considered to be witty and entertaining have built their reputation by responding in unexpected ways. Oscar Wilde, a master of wit, would typically remark, "I never put off until tomorrow what I can do the day after tomorrow."

Innovation breeds creativity. Therefore Ted's recipe can render someone more creative. It suffices to begin with a banal fact of life, state its opposite and then "massage" it until it makes sense. For example, "work is good for you." You can become an advocate of the *noble* merits of "not working." In ancient Greek society the fact that slaves rendered working for living unne-cessary made its citizens spend their time talking, philosophizing, and indulging in the arts and sports with such beneficial byproducts as the Parthenon, the Olympic Games, and democracy. Ted applied his recipe with much success some time later in Chandolin by turning around the well known motto "if you want, you can," to "if you can, you want."

Finally, it is ironic that if both an argument and its opposite can be used to prove a point, they can also be used to prove the contrary. For example, you may convince yourself, on the way to buy the newspaper, that you will win either way, whether there papers left or not. If there are still papers, because you'll get and read the paper, and if there no more papers left, because you can buy an ice-cream with the money instead. The devil's advo-cate could equally well argue that you will lose either way. If there no papers left, because you will not be able to read the paper, and if there are papers, because you will not have enough money for an ice-cream. Whenever you can argue that you'll win no matter what happens, someone else can always argue that you'll lose no matter what happens.

There have been so many studies carried out that for every study one can almost always find the anti-study, i.e., a study that supports the opposite. Ted knew he could argue successfully defending any position. His father had been a lawyer who had never lost a case. Ted, who inherited some of his father's genes, frequently debated with colleagues during lunch. The topics ranged from how good a movie or a theater play was, to the correctness of political decisions, the handling of criminality, the meaning of life, etc.

As he got older Ted reduced his participation in such discussions. He realized that he could pick *either* side of *any* debate and defend it *equally* well, only to finish half an hour later exhausted with the sole satisfaction of having played a good game; the content did not really matter. He did not worry now that his non-involvement might lead him to apathy. He slowly developed an appreciation for economy in the context of the word's original meaning: optimum dispensation. He decided to indulge in debates only to the extent that there was some concrete gain to be achieved and not simply for the sake of the discussion. He began to appreciate how silence can be golden.

20 – Life Feeds on Differences

The intricate relationship between opposites conceals more secrets. One day the speaker at the CERN seminar was not a physicist but a top executive from an electricity company. He talked about alternative sources of energy. He had prepared a little demonstration. He placed a small apparatus with two reservoirs in the auditorium's sink. (True to physics tradition the CERN auditorium, for all its sophistication, also featured a sink with running water next to the huge sliding green boards behind the podium.) He filled one reservoir with hot water and the other one with cold. In a matter of seconds a fan placed next to the podium started turning with electricity generated simply from the difference in temperature between the hot and cold water. The speaker's conclusion was a proposal to exploit the thermal difference of the ocean water (warmer at the top, colder deeper) as a source of energy.

The fact that energy can be harvested from any form of a potential difference was very familiar to Ted. Whenever he watched water fall at a waterfall he thought of the transformation of the water's potential energy at the top to its kinetic energy at the bottom. This kinetic energy can put a generator in motion and produce electricity. The difference in elevation between the beginning and the end of the waterfall constitutes a source of energy. In physics it is called a potential difference and applies just as well to differences in temperature.

Similarly differences in electric potential (like that between the two poles of a battery) produce electric currents that make electric appliances run. In fact *any* potential difference can serve as a driving force. Many examples can be found in society. The potential difference between those who have money and those who need money drives the economy. The economy would suffer in either case with everyone being very rich, or everyone being very poor. In the latter case because no one would be able to afford buying anything. In the former case because it would be impossible to hire poor people for the menial but necessary jobs such as collecting garbage, unloading trucks, and working in factories. Smart people would have no competitive advantage if there were no stupid ones around, and if all women looked like Demi Moore, she could not have asked for an exorbitant salary. If all lions ran equally fast and all gazelles ran equally fast, either the lions would starve to death (in the case gazelles outran the lions) or the gazelles would become extinct (in the case lions outran the gazelles). There would be no philanthropic organizations if everyone were equally happy (or miserable) and there would be less motivation to improve medicine if everyone were healthy or equally sick.

A potential difference can also be dynamic instead of static, for example variation over time of currency exchanges and stock prices. There could be no money to be made in the stock market if stock values did not change with time.

The larger the potential difference and the shorter the time between a swing from one extreme to the other, the stronger will be its impact. Ted thought that *life intensity* should be defined as the ratio: change over time. It is related to the mathematical derivative, like the case of speed. At twice the speed, there is twice as much new scenery coming into your field of vision, in the same unit of time. Similarly, excitement in the stock market will double in a certain day if the price change doubles.

Potential differences constitute sources of a vital force that drives and nourishes markets, stock markets, economies, competition, and survival. At the same time potential differences are consumable because their energy content can be harvested and used. When Ted confronted an Indian colleague with the fact that everyone's existence on Earth has a life cycle and therefore someday would be completed, the Indian physicist answered, "I am in part energy, and that cannot be destroyed!"

It is true that energy cannot be destroyed, or created, but it can be degraded. Degraded energy is less useful. For example, electricity is degraded into heat while using an electric iron. Once the heat becomes diffused over clothes and environment, it is impossible to recover it and use it again. Even though it has not been destroyed, it has become useless. The least degraded form of energy is sunlight; the most degraded form is ambient heat. Sunlight arrives on Earth and through successive transformations it takes progressively more degraded forms as it becomes stored in vegetation, charcoal, gasoline, electricity, etc. Life feeds on the intermediate stages—potential-difference steps—of this transformation chain.

Psychology acknowledges the beneficial aspect of swings between opposites. It points out that closeness and friendship can ensue from a good fight. It is true; making up after a fight is sweeter than if the fight had not taken place. Dostojewsky wrote that people might welcome illness for the sake of later enjoying getting better. One psychological explanation of why people rush to the scene of an accident is that they want to "enjoy" more intensely the fact that they are not involved in it. Happiness exists only when contrasted with unhappiness. If everyone were equally happy all the time, the phrase "I feel happy" would lose its meaning; it would be equivalent to saying "today is another day."

Society's increasing concern with diversity—be it biodiversity, cultural diversity, or other—is an indication that society is becoming aware of diversity's beneficial consequences. Diversity consists of many potential differences and as such is life-giving and should be preserved in all its forms. This has been amply recognized by environmentalists for what concerns biodiversity, but whereas biodiversity is threatened by pollution, cultural diversity is threatened by globalization.

Potential differences, variations, differentiation, and diversity bring richness to life. They enhance life intensity because the opposite—uniformity and lack of differentiation—resembles death. In fact, a visualization of the universe's ultimate death, given that energy (and matter) cannot be destroyed or created, is a universe in which all matter and all energy are uniformly distributed throughout space with no variation whatsoever. This state would not be void of energy. There would be much energy in it but it would be uniformly distributed with no potential differences and consequently useless. Hell could indeed be very hot, but could not have any hot or cold spots in it, because if it did, someone would be able to exploit the temperature difference and produce electricity to run an air conditioner!

In any case, this is not necessarily the end of the universe. It is possible that the universe will collapse into a singular point to undergo another Big Bang later thus entering oscillations of enormous amplitude and timeframe. Oscillations take place between potential-difference extremes and hence are life-giving. In this light, egalitarian social systems and notions such as communism, fraternity, and equality, if pushed to extreme, would lead to society's death. Capitalism, on the other hand, favors potential differences. But if the polarization between the haves and the have-nots grew to extreme, social unrest may ensue. The key to avoiding disaster in both cases seems to be the exercise of moderation.

The renowned American economist Joseph Schumpeter described capitalism as a process of "creative destruction." The contradiction in this characterization contains the element of the dynamic swing between potential and kinetic energy. An economic boom (kinetic energy) cannot occur if it is not preceded by a recession (growth potential). The wavy pattern of such a cycle is similar to a sine wave, the mathematical description of a pendulum's motion.

Ted knew well the harmonic oscillator. He was first acquainted with it as a teenager in high-school physics but he came back to it several times later with increasing sophistication in university. The climax though was in a Mendios meeting when Mihali used the harmonic oscillator to answer one of Ted's recurring questions concerning the exercise of moderation.

21 – The Harmony in Moderation

It had been Ted's turn to talk at that Mendios meeting and he decided to present harmonic motion in detail and map it on real-life non-scientific examples. He began by explaining how the pendulum's speed is maximum half way between its extreme positions where the process has the highest momentum and presents the highest resistance to changes, but the restoring force is zero.

"As we move away from the center position," Ted lectured, "the restoring force progressively increases (this force is proportional to the distance from the center point); it attains a maximum value at the excursion's extreme positions where the speed slows down to zero and changes direction."

Ted stopped and sketched a pendulum and its movement.

"As a pendulum swings between extreme positions," he continued, "it passes from a state where all its energy is in potential form (at the extremities where motion momentarily stops) to a state where all this potential energy becomes transformed to kinetic energy (at the middle of the trajectory), where the pendulum is moving at top speed. The speed of the pendulum oscillates between zero and top value regularly and periodically yielding a sinusoidal pattern a sine wave.

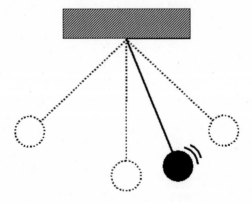

"This pattern features a continued procession of smooth highs and lows. This type of diversity over time is ideal, because it maximizes differentiation while it eliminates the risk of violent consequences. This is so because the motion of the pendulum is gifted with a special kind of smoothness. Operating with time on a sine wave as we did before—taking a derivative in mathematical terms—produces an identical sine wave. The absence of sharp edges from the original pattern results in a time-derivative pattern as smooth as the original pattern. This is an exceptional situation. As a rule, small anomalies encountered in real-life imperfections result in large consequences. For example, when in motion, a small but abrupt change of speed (say hitting a concrete wall at 10 miles per hour) can produce considerable damages. In such a crash, collapsible material reduces the damages by prolonging the duration of the impact; the change in speed is then less abrupt and the damages smaller. Operating with time means dividing by time and the most insignificant change, if it takes place in zero time, will release theoretically an infinite force with disastrous consequences. In contrast if the speed changes smoothly as in a sine-wave pattern, the forces released will follow a similarly smooth pattern. This symmetry between a pattern and its derivative may be the reason that the word 'harmony' has been associated with the pendulum's motion."

In his discourse Ted pointed out that harmonic oscillatory motion is typically encountered when there is an "overshoot" of an optimum level, a set target, and an established trend or fashion, necessitating corrective action to be taken. The change of direction can be anticipated whenever we see an excursion that seems extreme, for example, the swing of ex-communist countries to the right following the breakup of the Soviet Union. When the pressure builds beyond the break point, there is some kind of collapse and an overshoot in the opposite direction.

Ted then explained that overdoing anything results in oscillations. Such oscillations usually decrease in amplitude with time. Damping social mechanisms are invariably linked to cultural forces (culture acts as inertia). Long traditions result in "heavy" cultures that dampen social oscillations. This is the case with Europe compared to the US. Fads have higher-amplitude ups and downs in the US than they do in Europe. An extreme such case was the moon-landing adventure.

Visiting Earth's natural satellite can be considered as an American overreaction to Russia's launching of the Sputnik in 1957. While space exploration and exploitation is a valid endeavor, the moon landings did not justify the effort and the expense at the time, which is why Americans eventually discontinued the program. But in reaction to Sputnik Americans overshot and went to the moon. This overshoot may have contributed to the slowdown of the American space program in the 1980s. The slower-moving, more inert European space program saw smaller-amplitude fluctuations throughout its existence, and missed the moon adventure altogether.

At that point Ted raised the abstraction level of his discussion and pointed out that cultural diversity at world level (having low-inertia cultures as well as high-inertia cultures) made the world a richer place. He would have been less proud as a human being, he confessed, "had 'we' not gone to the moon."

It was at that moment that Mihali intervened.

"You can use harmonic motion to demystify the *pan-metron-ariston* dictum," he said.

One of the most popular samples of ancient Greek wisdom is encapsulated in the expression *pan metron ariston*, meaning to do everything with measure or in moderation. Ted heard this dictum time and again but always raised the same objection: what would constitute the correct size of the measure to be used? The dictum lacked precision. Ted needed a more objective rule to quantify the amount of moderation appropriate in every situation.

His objection was legitimate. Diversity and life intensity, as desirable as they may be, could reach extremes. It is possible to have too much of a good thing. Abrupt or excessive variation may produce violent consequences. For example, sudden large changes of temperature break glassware. Extreme polarization of the masses—as with the French and Russian monarchies—resulted in large-scale revolutions. And the opposite is also true. The smaller the differentiation, the lower the life intensity and the closer to the lifeless state of a completely non-differentiated environment with no potential differences whatsoever, the ultimate death. Moderation should be exercised at the right measure but what constitutes the right measure?

Mihali knew of Ted's stumbling block against *pan metron ariston* so he volunteered a way out.

"The pendulum swings around the middle vertical position, its average. But it also reaches both extremes," he said. "Exercising moderation with *harmony* does not mean to stubbornly stick to the average behavior but to swing between the extremes in a *harmonic* way. You can obtain quantitative guidelines on how to do that from the sine-wave pattern. It tells you where you should be as time goes on," and he sketched the familiar sine-wave pattern that characterizes both the position and the speed of rhe pendulum.

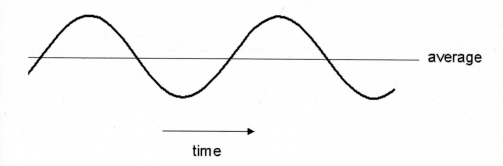

Ted listened and watched intently. "Notice," he remarked, "the pendulum spends most of its time away from the average position. It crosses the average line rapidly."

His comment prompted Mihali to continue, "Exactly! Take sex, for example. Let us say that young couples have sex once a day on the average. A naive interpretation of moderation along the lines of *pan metron ariston* would be to aim for this norm. But if you want to add harmony to your sex life, you should do rather the opposite, that is, alternate between days of intense sexual activity and days of no sex at all."

He then added as if in an afterthought, "The ancient Greek Epicurean philosophers—seekers of pleasure—did not advocate gluttonous gratification. On the contrary, they considered that a certain amount of refraining from worldly pleasures (austerity, abstention, fasting, etc.) was an indispensable ingredient for enhancing later pleasure. This is in sharp contrast to typical nouveau-riche behavior such as that of the Romans who feasted unrestrained to the point of having to induce vomiting in order to be able to continue eating. Elvis Presley may have eaten himself to death. He was poor as a youngster but later, when he could afford all the hamburgers in the world, he was unable to restrain his eating."

22 – The Symbolism of the Cross

Ted's discussions with Peter Brook in Amsterdam left such an impression on him that he sought the advice of Michel de Salzmann. When he had the opportunity he told Michel of his exchange with Peter and how he had advised him not to indulge in thinking any more but to put his energy into being, and of the large impact those words had on him. Michel listened carefully but seemed puzzled.

"I don't know what Peter had in mind," he said. "But one thing is clear, you cannot give up intellect. It is the only thing that distinguishes us from animals."

Then he walked over to his library, searched for a book and brought it to Ted. It was an old French paperback edition of a small book by René Guénon, *The Symbolism of the Cross*.

"Read this book," instructed Michel, "and then in Chandolin next summer you can talk to us about it. Not about the book," he clarified. "You talk to us about your own thoughts on the symbolism of the cross."

The book's archaic French rendered it unreadable for Ted. He also began suspecting that Michel did not understand him, whereas Peter Brook had. Still, he welcomed the idea of addressing the crowd in Chandolin again (he would receive quality attention!) But there was no way he could dig into this book for contextual material.

The Chandolin camp was scheduled for August, which gave Ted the opportunity to take his traditional vacation in Greece beforehand. He decided to solicit Mihali's help, as he was now living in Greece. "I need to talk to you," Ted told Mihali in a telephone conversation. "I have to give a talk on the symbolism of the cross this summer."

Ted knew that this was one of the many subjects on which Mihali had become an expert. Every time something triggered Mihali's fancy he went in great lengths studying, understanding, and elaborating on it. Ted remembered numerous drawings and 3-dimensinal models with which Mihali had filled every available space in his New York apartment. Each item had a story to tell invariably encoding some philosophical, anatomical, cosmological, or other secret. Aris had once remarked in New York City that Mihali could justifiably die, considering how much he had already achieved in his life.

"You are right! You need me," was Mihali's response on the telephone with no false modesty.

The occasion gave the two friends the opportunity to see each other outside Mendios and in an idyllic setting. They agreed to meet in a summer resort on the Sithonia peninsula in Chalkidiki across the bay from Mount Athos.

They had not seen each other for some time and again there was much to talk about. When they arrived at the beach resort they spent little time socializing. They wanted to be far from other people, so they could make all the noise they wanted. They gathered provisions for a long night out. They took food, drinks, a flashlight, paper and pencil, and materials to make a fire.

They walked to a remote spot of the beach and under the bright summer stars they built a fire on the sand. The dark mass of water in front of them ended with an imposing mountain range, Mount Athos, where there is the highest concentration of Eastern Orthodox monasteries in the world. How appropriate, thought Ted, to explore the symbolism of the cross in such a setting.

For the better part of the night Mihali did most of the talking, sometimes scribbling on the fine sand with a stick. Ted listened and took notes. By daybreak they stopped. Not because they ran out of things to say but because the sensation of the approaching sun engulfed them in a meditative atmosphere.

Ted now had much material with which to prepare his talk. He even had a recipe on how to construct a wooden model of a 3-dimensional cross and connect the tips of its arms with a tape thus revealing the mysterious symbol called Enneagram.

The Enneagram is a mystical symbol loaded with secret knowledge and occupies a central place in Gurdjieff's teachings. It encapsulates—for those who can interpret it—profound ancient secrets. At group meetings people rarely talked about it and if they did, it always was with seriousness and respect, as if they were talking about something sacred.

Some scholars attribute the origins of the Enneagram to the esoteric Christian teachings of the Desert Elders, the Fathers and Mothers of the Church, whose teachings have been followed in the monastic tradition of the Orthodox Church.

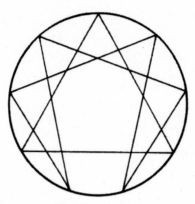

The Enneagram

Ted and Mihali bid each other goodbye the next evening. Ted was excited. He could see himself standing up in front of the Chandolin gathering in two weeks' time and presenting new ideas to people who were supposed to be experts in these subjects. Moreover, he would be doing it in the scientific tradition, that is, via observation, geometry, and logic, and not the way Guénon and so many others presented their theses, which mostly consisted of beliefs and subjective arguments too often biased by personal likings and wishful thinking.

Two weeks later Ted went to Chandolin. This time he carried some special luggage with him, a projector, a roll-up screen, a flip chart, and two wooden models of 3-dimensional crosses with thin red tape winding around them connecting the tips of their arms.

As he walked toward *Le Zoc* he paused looked at the imposing scenery around him and took a deep breath. The jagged mountain skyline across the valley demanded admiration; the air was thin and fresh.

At *Le Zoc* he now felt like an old-timer and even played host by giving directions to some newcomers. He knew where to go and what to expect. He directed himself to the small chalet where he chose his bed. He was becoming somewhat blasé about the whole thing but not about his upcoming presentation. It was scheduled for Wednesday evening after dinner.

During the next few days Ted went through the usual motions, but in his mind he'd neglect Michel's exercises in favor of rehearsals of what he was going to say.

Wednesday's dinner was low key and almost rushed. Michel had announced that Ted would be making a presentation and most people familiar with Ted's provocative style were eager with anticipation. As soon as dinner was finished people rearranged their cushions on the floor and Ted set up the projector, screen, and the flip chart. Finally he made the following introduction. He was speaking in French.

"I was asked to talk to you about the symbolism of the cross in Gurdjieff's work. But I can do it only in the languages I know, that is, broken French and physics."

There was some agitation. They were getting used to his poor French but were apprehensive about the science part. "Would you please speak slowly," asked José, Michel's wife, while someone murmured, "He is capable of talking faster than his shadow."

"The cross," began Ted, "is a geometric construction endowed with symmetry. To understand what it can symbolize we need to go back in time, much further back than the time of the early Christians. In fact we need to go all the way back to the beginning of the universe.

"One of the most fundamental laws of physics is the second law of thermodynamics, which stipulates that the entropy in the universe can only increase with time. Entropy can be thought of as a measure of disorder. So this law effectively says that the overall amount of disorder in the universe always increases even if occasionally order-increasing phenomena can be locally observed.

"If entropy is *always* increasing, it must have been minimal in the beginning of the universe and should become maximal at the end of the universe. In other words, at the time of the Big Bang there was utmost order in a super dense practically dimensionless and utterly symmetric state. In contrast, at the end of the universe there will be utmost disorder diffused everywhere, absolute chaos. In between, there is a progressive transition from order to chaos."

Ted drew a line on the flip chart and put down the family members that Mihali had assigned (political correctness was not an issue at that time.)

"The cross," continued Ted, "is a symbol of sufficient formal order to be close to the beginning. A snakelike irregular wiggle has low order and belongs closer to the end.

Let us look at such geometric forms as:

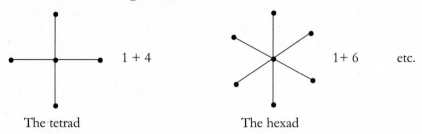

The tetrad   The hexad

"The complexity (order) of the cross can serve as a building block. Shapes like the tetrad, the hexad, and the like—the sons—held together by the dot in the middle (call it the monopole), have sufficient complexity to serve as building blocks for the forms of beings, hierarchies, species, families, and other complex entities.

"On the other hand, the absolute chaos at the end is characterized by extreme disorder and constitutes another polar extreme. It has no repetition, no periodicity, no auto-similarity, no spectrum, or other structural aspect. It cannot be incorporated as a component or a constituent of beings.

"My thesis is that all stable manifestations of entities, as diverse as light, atoms, humans, families, and societies, result from a balanced union between the son (cross) and the daughter (snake). The space between + and ~ on our line must be filled with symbols like these."

Ted had prepared a number of large cards with drawings on them. He showed the first one to the crowd.

"Many of these symbols occupy a prominent social or cultural position," he continued.

"But let us look at some examples from the sciences. Physics treats light as electromagnetic radiation in which the electric field E and the magnetic field H oscillate between positive and negative values. The two fields are orthogonal, that is, when E is large H is small, and as E gets smaller H gets larger. This pattern gives rise to a cross," Ted showed the following card:

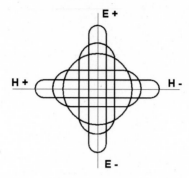

"But seen as a function of time the pattern traces 'snakes' bound in a 4 symmetry. Still, it is a stable and conservative pattern with no room for surprises," Ted showed another card.

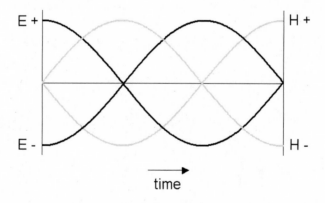

"Another example, poorly known in sociobiology where it really belongs, is sexual behavior in family and society. Imposing the cross structure on the freer sexuality of the snake results in the Ankh, which is known as the Egyptian symbol of fertility. What is not known about this symbol is that it holds the key to building stable and societal structures." Ted had a yet another card to show.

M biological mother
F biological father
S biological son
D biological daughter

"The black lines designate the well known sexual taboos. Sex is allowed between father and mother but not allowed with others outside this nucleus. There is a taboo between son and daughter, and there is a double taboo between parents and children, one for incest and one for the generation gap" he said and continued, "this set of taboos constitutes a conservative recipe because orderly sequences can now be formed as daughters and sons form couples and become mothers and fathers. Stable societies can be formed in this way."

There was a murmur of appreciation as Ted showed his next card that resembled the sketch of crowd.

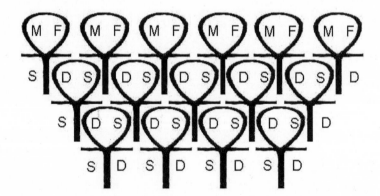

Ted went on, "We should be aware of the fact that the Ankh is also encountered in physiology. Many structures of the human form have this combination of cross and circle. A section of our spinal column looks like this:

and the setup of the entire nervous system is like this:

because there are interconnections in the head but not in the members.

"It is somewhat disconcerting that the Ankh seems to be etymologically related to the Greek word ΑΓΧΟΣ meaning anxiety. As if the latter was somehow intimately linked to the crux of society.

"The pattern of the cross also abounds in Chemistry. The carbon molecule is typically one octant of a 3-dimensional cross, the ideal tetrahedron:

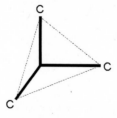

"Carbon becomes diamond through the action of fire and pressure. There is more density and more order in diamond than in Carbon. The figure of the 3-dimensional cross with all its points interconnected becomes the diamond shape and belongs further upstream in our cosmic time-entropy line.

"We can corroborate this chronology etymologically, something that cannot easily be attributed to pure coincidence. The Greek word for Carbon is ΑΝΘΡΑΞ, the root of which is the same as in ΑΝΘΡΩΠΟΣ, which means human being. But the diamond belongs further back in time. The Greek word for diamond is ΑΔΑΜΑΣ, the root of which is ΑΔΑΜ, which means Adam.

"We can push this etymology a little further by observing that in Greek the word ΑΔΑΜ is akin to the word ΑΔΜΗΤΟΣ, which in turn comes from ΑΤΜΗΤΟΣ, obviously related to ΑΤΟΜΟ (the atom as well as the individual) and ATMAN, the cosmic notion of universal soul encountered in Hinduism and Buddhism.

"Indisputably the diamond shape, rising from the 3-dimentional cross, belongs somewhere very early in the days of the universe."

Ted took a long breath here as he paused ahead of the final thrust of his presentation. The audience was listening as if hypnotized.

"If we attempt to connect the tips of the 3-dimensional cross with a continuous string in a unidirectional way, we will realize that there are only two ways of doing it. One is the Solomon way:

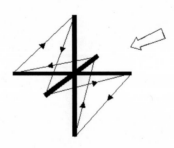

the projection of which into the indicated direction gives a regular hexagon."

Ted turned on the projector and held up the first wooden cross model. With minor adjustments the shadow cast on the screen was a regular hexagon.

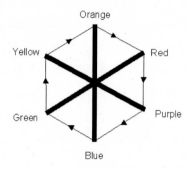

"Putting the six colors around the hexagon in the same sequence as they appear in the rainbow gives complementary colors on the extremities of each axis (complementary colors are colors that produce a neutral gray when mixed together.) The resulting circulating flow seen from the side gives again rise to a cross pattern (the axis around which the whole thing circulates and the plane). One could say that the relationship of God (axis) and Man/Organic life (plane) is orderly and conservative because it begins with the cross and gives back the cross.

"The other way of interconnecting the six points of the 3-dimensional cross in a continuous way gives the Gurdjieff way:

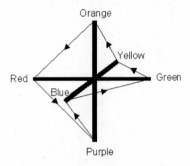

"Putting again the six colors in their sequence in this configuration gives complementary colors on the extremities of only one axis (in this example red-green). But it is possible to define a new direction in such a way that the other four colors find themselves each one vis-à-vis its complementary color *with respect to this new axis*. The way to do this is to project the whole thing into a plane perpendicular to this new direction."

Ted held up the second wooden cross model in the projector's light. Observing the shadow it cast on the screen he struggled for a while trying to orient it properly. Finally the shadow on the screen revealed the fundamental pattern of the Enneagram. Another muttering arose from the audience. Ted felt as if he was divulging some of the Enneagram's secrets.

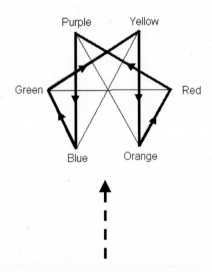

"With respect to the new axis," Ted pointed at the direction of the broken-line arrow, "each color faces its complementary color. Moreover, the circulating flow now—the relationship of God (axis) to Man/Organic life (circulating string)—looks like this:

"There are similarities between the Solomon and the Gurdjieff way, but the latter is more intricate, has more of a 'snake' quality, and therefore is more advanced, i.e., later in time (greater in entropy/disorder)."

• • •

Mihali had given the biblical name "Solomon" to the first configuration of connecting the six points of the 3-dimensional cross and Ted had kept it. It seemed appropriate because the configuration was symmetric, rational, and straightforward. Also, Solomon preceded Gurdjieff.

• • •

Ted continued, "Besides science, alchemy has also made use of the cross pattern. Alchemists have placed the four elements on the points of the cross:

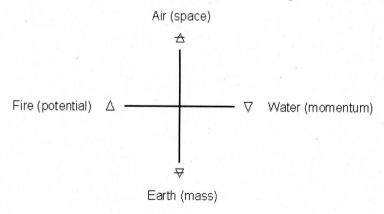

In parentheses we see the physical-science analogs.

In a pictorial representation of a human being one could see Intellect (air), Body (earth), Will Power (fire), and Love (water). The third axis in the 3-dimensional cross (not shown on this figure) can represent the dimension reception-projection or perception-expression, aspects that map to the masculine and feminine archetypes.

Gurdjieff's work can be seen as a reconciliation of opposites, in other words, a harmonious development of human beings so that they occupy a place close to the center of the cross. The aim is to avoid being outcast in an extreme, be it love or will power; one should not live by intellect alone nor cultivate only the body. Guided by the cross we are striving for the harmonious admixture of the four elements, thus gaining order and beating entropy, if only exceptionally.

I will leave you with a question," Ted prepared to conclude.

"Such an anti-entropic evolution can take place only locally. In other words, order can be increased somewhere at the expense of increasing disorder (by at least as much) elsewhere. Where is then to be found the disorder that corresponds to the internal order we are achieving here via the Work?

"Could it be that this is one way to understand Gurdjieff's provocative claim that knowledge is a limited resource, and therefore collecting it somewhere results in depleting it elsewhere?"

Ted stopped talking, stood quiet for a short while, and then he sat down. No one spoke for a long time. Ted had touched and manipulated concepts that people in Gurdjief groups respectfully avoid. All heads were turned toward Michel with expectation. Finally Michel spoke.

"Where does all this information come from? What are your sources?" he asked calmly.

Ted stood up and articulated clearly. "Mihali Yannopoulos," he said and sat down.

There was another long silence. There must have been questions in everyone's mind about who this person may be, but no one dared asked. Ted did not volunteer any more details either. He knew that Michel had received Mihali's 30-page letter asking to be admitted to a camp at Chandolin. Ted was now convinced that next summer Mihali would be able to attend.

23 – Mind over Body

That winter was rather stressful for Ted. His contract at CERN was coming to an end and there were no viable alternatives in sight. He began exploring possibilities for employment in less competitive academic positions around Europe. But these alternatives, besides the stigma expressed in a saying by physicists "if you can't do research, you can always teach," also involved a relocation for at least Ted if not his whole family. At this time attending routine meetings at the Gurdjieff group at the rate of two per week became somewhat of a burden. There was only one new distraction for Ted: skiing.

He had tried skiing several times before but always unsuccessfully. For several years, once a year, he had been persuaded to join colleagues from CERN on a skiing outing. Invariably the experience turned sour for him. He wasn't much of a sportsman and his Swiss colleagues believed that skiing comes to you naturally the first time you put your feet on skis. On every occasion Ted came back bruised both physically and psychologically, and finally swore never to go skiing again. But this January something happened that made Ted change his mind.

He went to a physics conference held at a ski resort in the French Alps. The presentations were all scheduled from 8 to 10 am and 4 to 6 pm. The rest of the day was free. The conference lasted for a week. Ted watched without much regret the other physicists become excited about skiing every day. But on the last day he gave in. A Japanese theoretical physicist, older than Ted and speaking no French and hardly any English, wanted to take a skiing lesson and asked Ted to come along as a translator. Ted decided to rent a short pair of skis for himself and take the lesson too. The lesson lasted half an hour. A young lady ski monitor showed the two men a few things and sent them off saying, "Now you are ready to go!"

The two men got on a chairlift and went up the mountain. On the way up the Japanese physicist asked Ted, "What she said? Put weight on left foot or on right foot?"

Ted laughed; he could not believe the man's ignorance and naïveté.

"It depends which leg is downstream," he said. "You always put your weight on the foot of the downstream leg."

At the end of the chairlift they both managed to come off the chair without falling. But before Ted had a chance to remark on their awk-wardness the other man headed straight for the downhill run. Ted rushed to follow him so as not to lose him. Struggling not to fall, Ted mumbled to himself, "This man is older than I and more ignorant. I'll be darned if he leaves me behind."

It was not a very long ski slope and they both reached the bottom of the mountain without falling. They were beaming with excitement. They looked at each other and exchanged:

"Again?"

"Again!"

They went up the mountain and down again the same way. Ted was elated.

"I am going to learn how to ski," he vowed.

And this is how Ted became fascinated with skiing. The experience of skiing down the slopes of a high mountain speeding effortless and quietly while admiring a breathtaking view proved exhilarating.

Back in Geneva he decided to teach his 10-year-old son, to ski. The boy was a quick learner and soon the two of them were going down difficult slopes. One day they ended up on a black-coded run. The steepness was frightening and the boy became demoralized and began crying. Ted encouraged him, himself taking risks. "There is nothing to it. Look how easy it is to turn and to stop," he would say, demonstrating it to the best of his ability.

Getting down the slope proved long and painful, particularly for the youngster. "I don't want to come here again," he said afterward. And yet, a year later, the two of them, plus a friend of his son's, found themselves on the same hard slope. This time the two kids were ahead and Ted struggled to follow. "Come on, Dad, we are waiting for you!" was the remark that seemed to make it more difficult for Ted. He fell a couple of times; he could not understand how he had gone down this very slope so effortlessly the year before.

Then he realized the importance that his mental state had over his body. When he felt that he was needed, he performed exceptionally well. When the others were better than him and he felt like a dead weight, he unwillingly behaved like a dead weight! He later found out that this is the leader's legs syndrome, known among military people; the fact that when you are in front, leading, you have an extra level of strength and stamina.

On several occasions Ted had observed impressive manifestations of the influence his psychic state had over his body. Often it was during the "movements" at the Gurdjieff group. In one such exercise they were asked to remain for a long time with their hands stretched out in the form of a cross. Less than five minutes into this exercise the fatigue became painful and the continuation of the exercise seemed unbearable. At that moment they were told to begin counting in loud voice as follows: one, two, three, four, two, three, four, five, three, four, five, six, four, five, six, seven, etc., and then backward: ten, nine, eight, seven, nine, eight, seven, six, eight, seven six, five, seven, six, five, four, etc. It turned out that concentrating on the count made them forget the fatigue and the pain; they were able to carry on with their arms stretched for a very long time beyond the point when they had felt they could not continue.

In another exercise they were asked to shout loudly while performing a march. The effect was invigorating and put each one of them into an excited and exuberant emotional state. Ted understood why the military employs such practices. Emotional energy amplifies physical achievement. Performers who face exceptional demands know this, if subconsciously, and they adamantly solicit the subjugation of emotional energy in various ways. The Work in Gurdjieff groups was supposed to do that, but it required years of preparation. There are quick-and-dirty alternatives to bring out the emotions like alcohol, marijuana, or other drug. Ted was delighted by a scene in a movie in which the father goes home after having had a few drinks at the bar, sits at the table and tells wife and son how much he loves them. When the son wonders at his father's unusual behavior, his mother says, "Don't pay attention to him, he is drunk." But then the son then wants to know, "Does that mean that when he is not drunk he does not love us?"

24 – If I can, I want

The following summer at Chandolin both Aris and Mihali were present. Furthermore Aris, who leveraged Ted's acquaintance to win acceptance, had brought two other Greeks with him and presented himself as the head of the Gurdjief group in Athens, and as such he claimed a place at the table's head next to Michel and the other elders. While Aris cultivated his relationship with the elders, Mihali concentrated in observing, living in depth, and experiencing fully every minute. Ted was thrilled to finally be in such an alpine setting with his old friends; otherwise the camp activities *per se* had lost their luster for him.

He expressed that feeling in front of everyone over dinner. He said that he no longer felt the enthusiasm he had experienced the first time he had come to Chandolin. Michel responded promptly and sternly.

"Why seek enthusiasm?" he said. "Enthusiasm is tantamount to cheap thrills. They consume your energy and distract you from your main aim, which is the Work. The emotional energy associated with enthusiasm is of lesser quality, like that with alcohol and drugs. You shouldn't run after that. The real emotional energy will emerge in you as a consequence of having achieved personal development."

Ted did not know at the time that it was his last camp at Chandolin. On the contrary, he went along as usual and in fact meticulously prepared a spicy story to tell during the Saturday-night fete.

Saturday afternoon at the sauna, Aris, Mihali, Ted, and two others, all naked, had a dirty-joke contest. Aris's joke was the best.

> A lion wandered alone in the jungle feeling lonely. All others animals afraid of him kept their distance. When he saw a little monkey up on a tree the lion asked him to come down and keep him company.
>
> "No," said the monkey, "I am afraid of your claws."
>
> "Don't be afraid," reassured him the lion, "Here, I'll tie my hands with a rope," and proceeded to tie up his front legs with a rope.
>
> "Will you come down now?"
>
> "No," insisted the monkey, "because you also have claws in your hind legs."
>
> "OK," said the lion, "I'll tie those too," and he tied up his hind legs as well.

"Will you come down now?"

"No," continued the monkey. "Your big teeth are scary."

The lion became annoyed. "I will tie my mouth too," he said and tied up his muzzle the best he could. "Will you come down now?" He mumbled through the ropes.

The monkey climbed down the tree slowly but hesitated and trembled as he approached.

The lion got impatient, "What is your problem? My hands are tied, my legs are tied, my mouth is tied; what the hell are you afraid of?"

"I am not afraid," replied the monkey. "I am simply nervous because it is the first time I will screw a lion."

Ted dressed up for the fete that evening. He wore a white turtleneck silk shirt with an old-fashioned sleeveless vest and striped pants. He chained his father's old watch at the vest's pocket with the chain hanging visibly attached to one of the vest's buttonholes.

Before entering the grand hall Ted and Aris went for a little walk in the sloping fields around *Le Zoc*.

"Tonight I am going to play the devil's advocate," confided Ted, his voice revealing a certain apprehension.

"I will support you," offered Aris, "if you promise me not to move away Geneva."

Aris knew of Ted's efforts to find employment and did not want to see him abandon the Geneva area. It wasn't entirely selflessly; Aris wanted to be able to visit Ted and use his place as a base for reaching central Europe.

"It's a deal," replied Ted, who had just about decided not to seek employment away from Geneva anyway.

Having concluded their pact, they entered the large dining room, specially arranged for the traditional end-of-camp fete.

Gurdjieff liked to drink the French cognac Armagnac and therefore during such fetes there was invariably a little glass of Armagnac at each person's place. It was drunk drop by drop because it was supposed to last through all the numerous toasts that were made during the festive dinner.

This time one of the toasts came from Aris. He seemed anxious to offer his toast early.

"I raise my glass to the forces that unite us together here, even if these forces come from the devil himself," he said glancing smilingly at Ted.

People raised their little glasses but as they drank, Ted choked loudly. All heads turned toward him. He was embarrassed; Michel also seemed somewhat disturbed but said nothing. The tension dissipated, more toasts were made, and stories were told. Once dinner was finished, Ted said in a slow imposing voice.

"I too have a story to tell, but mine is not a joke. I want to describe to you my real encounter with the devil," said Ted, and paused dramatically.

There was silence in the audience again because the mention of devil carried special weight in these circles. The notion of devil did not necessary carry sinister connotations. They considered the devil as being sly, and as such the role was often more envied than criticized. Gurdjieff's masterwork, a three-volume fictional novel-like philosophical treatise, is entitled *All and Everything* and has the endearing subtitle *Beelzebub's Tales to His Grandson.*

"But before I tell you about my encounter," continued Ted, "I want to tell you about a natural law that I have discovered and christened with the deserving Latin name *Dum Possum Volo*, which means if I can, I want. This law stems from a mathematical principle, the ergodic hypothesis, and the corresponding physics theorem. It says in words that if I can do something there will be forces that will push me to do it."

"Are you referring to 'where there is a will there is away?'" interrupted José, Michel's wife.

"No, madam," replied Ted, "I am referring to the opposite. Where there is a way, there is a will."

José looked thoughtful and stopped interrupting him, so he proceeded.

"We see manifestations of this law around us all the time. When George Mallory responded with 'Because it is there' to the question 'why did you climb Mount Everest?' he failed to add 'and I could.' Mount Everest is there for me too, but I cannot. The fact that he could was a crucial ingredient in the realization of that feat.

"When parents are asked 'why did you have children,' they may respond with 'because we did not want to miss the experience' or with 'because we love children' or with other more or less sophisticated statements. It does not change the fact that couples have children mainly because they can. Most of them do not even ask the question until after the fact, and only then try to find justification for it.

"What one can do changes with age and so does the desire to do it. I had a student who smoked two and a half packs of cigarettes a day. Being an ex-smoker myself, I suggested that he quit. 'I am still young,' was the reply. It becomes easier to stop smoking as one gets older. This happens not because age enhances willpower, but rather because the harmful physical effects of tobacco are less tolerated by a weakened organism. In other words, older people are less able to smoke. At the same time, they do more of what they can do.

"But let us come back to my encounter with the devil. As you know in our group in Geneva we devote one entire Sunday every month to the Work under Michel's supervision. On such a Sunday morning last winter I woke up and looked out my window. It was a bright, clear day with lots of fresh snow on the surrounding mountains. Ideal ski conditions, I thought. There are few of these days during the season and I liked skiing. In fact, I had just bought ski equipment that incorporated the latest technology. So I paused torn between my urge to go skiing and my decision to devote this Sunday to the Work.

"At that moment I heard a voice ask, 'Theodoraki, do you know the law *Dum Possum Volo*?'

"I immediately knew there was something strange because no one ever called me with that name other than my mother when I was a little boy.

"'Do I know it?' I said, 'I am its godfather!'

"'Very well then,' said the voice. 'Why are you hesitating? Go skiing, because you can. There won't be many days like this. You won't be young and capable of skiing forever. Take advantage of it now while it is possible. You can always go back to spiritual work when you are old and unable to do the things that you can do now.

"'I will tell you one more secret,' the voice continued, 'but don't tell it to anyone lest you get in trouble.'

"I have not told anyone up to now," confessed Ted, "but I trust you people so I dare repeat it here.

"The voice continued by saying, 'you will be taking no risks by staying away from your esoteric work today. You can always catch up with it later. When you die, whether you have worked for one year or for twenty years makes no difference whatsoever.' He said no more."

At this point Ted stopped and waited. It was again José who interrupted the silence.

"So what happened?" she asked.

"I had no chance," replied Ted, "the devil had done his job too well. Soon afterwards, I received a phone call telling me that Michel, who was due to come from Paris and direct the day's activities, was held up, but that we were supposed to go and carry on without him anyway. I went skiing that day."

Once Ted stopped talking the intensity defused but some uneasiness remained in the air. No one thought the story was funny. A little later, as people moved around rearranging the room, Aris approached Ted apprehensively.

"Have you finished, or do you still have more to say?"

"I am finished," replied Ted. Aris seemed relieved. He felt that so far he had kept his part of the deal to support Ted and was dreading the possibility of having to align himself with further seditious behavior endangering his newly acquired status among the elders.

On his part, Ted felt he had laid down his terms. He had effectively demanded and granted himself permission to attend group work selectively, i.e., only when Michel would be there and when there was no other competition for his time. Moreover, he had formally launched the *Dum Possum Volo* law and he had the premonition that it was going to play an important role in his life.

25 – Management Science

Ted and his wife had socialized with other foreign couples with young children. One of them was Nancy and John. Nancy was an African American sociologist from North Carolina who enjoyed singing Negro spirituals. John was an English-mother-tongue blond chemist of French descent. He had taught at the University of Geneva for a while. But they both finally ended up working for Digital Equipment Corporation. At that time DEC, as the computer firm was usually called, was at its zenith selling its popular mini-computers.

In one of their social get-togethers Nancy mentioned to Ted that DEC was hiring scientists for research work. Ted had never taken the industry option as a viable alternative for his job search, as physicists looked down upon non-academic nine-to-five jobs. There was a stigma involved in such a move, as if someone would sell his or her soul to the devil. Ted in particular felt he had been betrothed to science since early childhood, and despite his disenchantment he could not conceive of a "divorce."

But Nancy made a number of convincing arguments: he would be in a group of scientists, he would be doing research-type of work, and he would enjoy a high salary and a long-term stability without having to move away from the Geneva area. Moreover, John told Ted that his transition from the University's chemistry department had not only been painless but also enjoyable. Ted was convinced and made an appointment with Walter Meyer, the head of the Management Science group at DEC's headquarters. He turned out to be another particle physicist who had left CERN ten year earlier and became the founder of DEC's Management Science group.

Physicists had long shied away from industry and disciplines they considered to be pseudo-sciences. But finally, squeezed by diminishing returns, many physicists like Ted and Walter began to scatter outside the musty dungeons of mammoth particle accelerators. They were not alone; others had already shown up in unlikely places: neurophysiology, Wall Street, chaos studies, and forecasting, to name a few. They carried with them detectors, tools, techniques, and tricks, all built around the scientific method. They strived for understanding the future in the physics tradition—like hound dogs, sniffing out the right direction as they move along.

The management science group consisted of a dozen Ph.D.s: mathematicians, chemists, computer science specialists, a physicist, and a geologist.

They studied a variety of topics often of their own choosing. The general loose requirement was that their work should eventually benefit the company in some way. Ted would receive an 80 percent increase in salary.

Ted took the plunge. He did not take time to analyze the importance or the consequences of his decision. It was not the first time in his life that he marched along and made an important decision impassively like an outside observer. He remembered other such occasions, e.g., leaving America and handing over his coveted green card, getting married, and watching his children being born in the delivery room. In every occasion he had simply proceeded to do what had to be done without much deliberation, fear, or passion. As if he were obeying orders rather than setting his own course. He liked doing things this way because he felt that he followed a natural evolution. He whimsically referred to it as "rejoice rejoice we have no choice." At the same time, he did not feel that he betrayed Gurdjieff's command to strive for conscious acting. Ted was not behaving as "asleep" or "hypnotized," which is what Gurdjieff criticized in most people's everyday behavior. Ted was aware that some natural process was taking place and he consciously let himself being carried along without resistance. That's how he defined volition: to be in harmony with what is going on and let it happen.

At DEC Ted was given the task of studying competition in the computer market. Many new computer models were making their appearance in the market, substituting for one another. The aim of the exercise was to eventually make forecasts of demand and sales. He found the project challenging and exciting. Foretelling what is going to happen carries magical overtones. It implies acquiring power, becoming super-human. Of course he knew nothing about competition or forecasting techniques. But he knew that the science "game" is not very different. From the observation of past events, scientists formulate laws that, when verified, enable them to predict future outcomes. Science revolves around methodologies for telling the future despite the fact that scientists normally disdain the notion of fortune-telling.

Scientists merely use a different vocabulary. In contrast to fortune-tellers, they talk about calculations instead of predictions, laws instead of fate, and statistical fluctuations instead of accidents. Formal forecasting endeavors by scientists have been undertaken in meteorology, economics, and the stock market. But despite formidable efforts none of them proved successful concerning the long-term future.

Ted's first reaction was to go to the biology library at the university and look up books and other publications on competition. He was impressed with what he found. There was a tremendous amount of rigorous mathematical work on the subject. The laws of survival of the fittest and natural growth in competition had been extensively studied. Interactions such as that between predator and prey, and between parasite and host had been cast into rigorous mathematical equations. Growth curves for animal populations follow patterns similar to those for product sales. Could it be that the mathematics developed by ecologists for the growth of a rabbit population describes equally well the growth of cars and computers? Ted would not hesitate to mobilize this science outside the realm it had been designed for. After all the scientific method is of universal validity and competition in the marketplace is not different from that in the jungle. The law of the survival of the fittest becomes indisputable.

He also found a number of interesting and provocative articles by Cesare Marchetti, who was a physicist at the International Institute of Advanced Systems Analysis (IIASA) near Vienna. Marchetti had been working on forecasting energy demands. He had also used the scientific method: observation, prediction, and verification. Predictions must be related to observations through a theory resting on hypotheses. When the predictions are verified, the hypotheses become laws. The simpler a law, the more fundamental it is and the wider its range of applications.

Marchetti had long been concerned with the "science" of predictions. In his work, he first started searching for what physicists called *invariants*. These are constants universally valid and manifested through indicators that do not change over time. He believed that such indicators represent some kind of equilibrium even if one is not dealing with physics but with human activities instead. He then assumed that the fundamental laws that govern growth and competition among species also describe human activities. He went on to make a dazzling array of predictions, including forecasts of future energy demands, using mathematical equations developed by biologists. But how far can the analogy between natural laws and human activities be pushed, and how trustworthy are the quantitative forecasts based on such formulations?

Ted became enthusiastic about Marchetti's work and studied it in depth. When a few months later he was asked to forecast the life cycles of computer products and the rate at which recent models substitute for older ones, his reaction was immediate. The first morning after he was officially charged with the forecasting project he took an early plane for Vienna to see Marchetti at IIASA. This would not have been possible in an academic institution, where there is much bureaucracy and trips must be planned well in advance. Ted found working in industry to be enjoyably dynamic.

26 – The S-Shaped Adventure

Ted arrived at Laxenburg, Austria in 1985, seeking out Marchetti at IIASA. Laxenburg is a tiny old town that has stayed in the past even more than the nearby capital, Vienna. IIASA is housed in a chateau with a fairy-tale courtyard in which Marchetti frequently strolled. His silver hair, spectacles, Tyrolean feathered hat, and Austrian cape made him the classic image of a venerated Austrian professor.

Not at all! An Italian, he received his doctorate in physics from the University of Pisa, and his activities evolved in a Leonardo da Vinci tradition to cover a wide area both in subject matter and geography. But he also enjoyed playing the devil's advocate, something that made him even more fascinating in Ted's eyes.

In 1972, a book appeared under the title *The Limits to Growth*, published by the Club of Rome, an informal international association with about seventy members of twenty-five nationalities. Scientists, educators, economists, humanists, and industrialists, they were all united by their conviction that the major problems of mankind are too complex to be tackled by traditional institutions and policies. Their book drew alarming conclusions concerning earth's rampant overpopulation and the depletion of raw materials and primary energy sources. Its message delivered a shock and contributed to the "think small" cultural wave of the 1970s.

Marchetti had not joined the Club of Rome because he considered that it had failed to stay close to the fundamentals of science. His response to *The Limits to Growth* was an article titled "On 10^{12}: A Check on Earth Carrying Capacity for Man," written for his "friends of the Club of Rome." In this article he provided calculations demonstrating that it is possible to sustain one trillion people on the earth without exhausting any basic resource, including the environment! It served as one more brushstroke in his self-portrait as a maverick.

He received Ted in his office buried behind piles of paper. Later Ted discovered that most of the documents contained numbers, the data sources serving Marchetti very much as sketchbooks serve artists. Marchetti welcomed Ted warmly and uncovered a chair for him. They went straight to the point. Ted showed him his first attempts at determining the life cycles of computers. Ted had dozens of questions. Marchetti answered laconically, simply indicating the direction in which Ted should go to search for his own conclusions. "Look at all computers together," he said. "Small and big ones are all competing in the same market. They are filling the same niche. You must study how they substitute for one another. I've seen cars, trains, and other human creations go through this process."

What he said produced an echo in Ted. People can spend their money only once, on one computer or on another one. Bringing a new computer model to market depresses the sales of an older model. Also, one can only type on one keyboard at a time. A new keyboard with ergonomic design will eventually replace all flat keyboards in use. This process of substitution is a phenomenon similar to the growth of an animal population and follows the fundamental law of nature described by ecologists.

Ted's discussion with Marchetti lasted for hours. Marchetti defended his arguments in a dogmatic way that often sounded arrogant. Some of the things he said were so provocative that it was difficult for Ted not to argue. However, Ted mostly kept quiet, trying to absorb as much as possible. During lunch Marchetti kept tossing out invariants as if to add spice to their meal. Did Ted know that human beings around the world are happiest when they are on the move for an average of about seventy minutes per day? Prolonged deviation from this norm is met with discomfort, unpleasantness, and rejection. To obscure the fact that one is moving for longer periods, trains feature games, reading lounges, bar parlors, and other pastime activities. Airlines show movies during long flights. On the other hand, restriction of movement is equally objectionable. Confinement makes prisoners pace their cells in order to meet their daily quota of travel time.

Did Ted know, Marchetti asked, that during these seventy minutes of travel time, people like to spend no more and no less than 15 percent of their income on the means of travel? To translate this into biological terms, one must think of income as the social equivalent for energy. And did Ted know that these two conditions are satisfied in such a way as to maximize the distance? Poor people walk, those better off drive, and the rich fly. From African Zulus to sophisticated New Yorkers, they are all trying to get as far as possible *within the seventy minutes and the 15 percent budget allocation*. Affluence and success result in a bigger radius of action. Jets did not shorten travel time; they simply increased the distance traveled.

Maximizing range, Marchetti said, is one of the fundamental things that all organisms have been striving for from the most primitive forms to mankind. Expanding in space as far as possible is what reproducing unicellular amoebas are after, as well as what the conquest of the West and space explorations were all about. In his opinion, every other rationalization tends to be poetry.

On the return flight Ted experienced a euphoric impatience. He took out his pocket notebook to organize his thoughts and jot down impressions, conclusions, and action plans. He decided to go straight to his office upon his arrival late Friday night. He had to try the predator-prey equations to track the substitution of computers. Marchetti had said that looking at the whole market niche would reveal the detailed competition between the various models. Could this approach describe for Ted the substitution of computers as well as it described the many substitutions Marchetti spoke of? Could it be that replacing large computers by smaller ones is a *natural* process that can be quantified and projected far into the future? Would that be a means of forecasting the future of DEC? Could the life cycles of organizations be predicted like those of organisms, and if so, with what accuracy? Would it even be possible to derive an equation for himself and estimate the time of his own death?

Beyond his excitement, Ted was suspicious. He did not know how much he could trust Marchetti. He had to check things out for himself. If there was a catch, it should become obvious sooner or later. The one thing he did trust was the scientific method.

In a few months most of Ted's friends and acquaintances knew of his preoccupation with using formulations by biologists and ecologists as a means of probing the future. Contrary to his experience as a particle physicist, interest in his work now was genuine. It was no longer the befuddlement of those who believe themselves intellectually inferior and admire the interest that must be there in something too difficult to understand. Now, it was more like, "This is really interesting! Can I make it work for *me?*"

Everyone who knew of Ted's work was intrigued with the possibility that forecasts could become more reliable through the use of natural sciences. Requests for more information, explanations, and specific applications kept pouring in from all directions. A friend who had just begun selling small sailboats wanted to know next summer's demand on Lake Geneva. Another, in the restaurant business, worried about his diminishing clientele and was concerned that his cuisine was too specialized and expensive compared to his competition. A depressed young woman was anxious to know when her next love affair would be, and a doctor who had had nine kidney stones in fifteen years wanted to know if and when there would be an end to this painful procession.

Besides those eager to believe in the discovery of a miraculous future-predicting apparatus, there were also the skeptics. They included those who mistrusted successful forecasts as carefully chosen successes among many failures; those who argued that if there was a method for foretelling the future, one would not talk about it but get rich on it instead; and those who believed that their future could not be predicted by third parties because it lay squarely in their own hands.

For his part, Ted soon reached a stage of agonizing indecision. On one hand, Marchetti's approach appealed to him. The scientific method and the use of biological studies on natural-growth processes and competition inspired confidence. On the other hand, he needed to make his own checks and evaluate the size of the expected uncertainties. He also had to watch out for the trap of losing focus by indulging in the mathematics. But if in the end he confirmed an increased capability to forecast, how would he accommodate the predeterminism that it implied? The belief that one is able to shape one's own future clashed with the fascination of discovering that society has its own ways that can be quantified and projected reliably into the future. The old question of free will would not be easily resolved.

Ted searched for more and more cases in which social growth processes fit the description of natural and biological ones. He also searched for discrepancies and carried out extensive simulations to understand the failures. He had embarked upon what he came to call his S-shaped adventure, and it took him in surprising directions. Along the way he became convinced that there is a wisdom accessible to everyone in some of the scientific formulations that describe natural-growth processes, and when they are applied to social phenomena, they make it possible to interpret and understand the past as well as forecast the future. Furthermore, he learned to visualize most social processes through their life cycles without resorting to mathematics. Such a visualization, he found, offered new perspectives on both the past and the future.

As he dug deep into the subject Ted came to two realizations. The first one dealt with the fact that many phenomena go through a life cycle: birth, growth, maturity, decline, and death. Time scales vary, and some phenomena may look like revolutions while others may look like natural evolutions. The element in common is the way in which the growth takes place; for example, things come to an end slowly and continuously, not unlike the way they came into existence. The end of a life cycle, however, does not mean a return to the beginning. The phases of natural growth proceed along S-shaped curves, cascading from one to the next, in a pattern that reinvokes much of the past but leads to a higher level.

Ted's second realization concerned predictability. There is a promise implicit in a process of natural growth, which is guaranteed by nature: The growth cycle will not stop halfway through. Whenever Ted would come across a fair fraction of a growth process, be it in nature, society, business, or his private life, he tried to visualize the full cycle. If he had the firsts half as given, he could predict the future; if he was faced with the second half, he could deduce the past. Ted succeeded in making peace with his arrogance and grew to accept that a certain amount of predetermination is associated with natural processes, as if by definition from the word natural.

27 – Using S-Curves to Unearth Secrets

On his next trip to Greece to see his ailing mother Ted's air ticket included a four-hour stopover in Athens. He wrote to Mihali and Aris to let them know. They could hold one of their meetings near the airport.

All these years since the New York days their correspondence had been rare but of a singular character. Whoever wrote a letter made two photo-copies and mailed them to the two others, keeping the original for his records. Each letter had to be answered (at least by an acknowledgement) before the same person could write another letter. And the letters were generally long. Mihali had once written a 50-page letter. But now there was a special reason that Mihali and Aris wanted to talk to Ted in person. Ever since they learned about his new activities they wanted to discuss the subject in depth, the way they had been accustomed to in the Order of Mendios.

Coming out of the arrivals building at the Athens airport, Ted saw the two hefty, handsome, and bearded figures jumping up and down jubilantly to draw his attention, as if otherwise he would miss them. They drove to a hotel not far from the airport, stormed in with briefcases and luggage, and asked for a well-lit quiet room with electrical outlets. The hotel owner said he had one, and somewhat intimidated, asked what this was all about. Aris took the hotel owner aside and very seriously remarked, pointing at Ted, "This gentleman comes from abroad. He is a scientist. We must talk."

They set up in the room with a tape-recorder, microphone, calculators and notepads. This was not going to be a usual-type Mendios meeting. They plunged into a heated three-hour discussion on the capabilities of foretelling the future via biological models and other scientific notions such as inva-riants and homeostasis. Ted presented one idea after the other. He soon came to the topic of associating the evolution of a person's creativity and productivity with natural growth. The work of art or science can be thought of as the final expression of a "pulse of action" that originates somewhere in the depths of the brain and works its way through all the intermediate stages to produce a creation. The number of creations over time grows along S-shaped patterns (or S-curves.) Most people die close to having realized their creative potential, the ceiling of their S-curve. The idea is intriguing. Obviously people's productivity increases and decreases with time. Youngsters cannot produce much because they have to learn first. Old people may become exhausted of ideas, energy, and motivation. It makes intuitive sense that one's productivity goes through a cycle over one's lifetime, slowing down as it approaches the end. The cumulative productivity—the total number of one's works—could very well look like an S-curve over time. But the possibility of formulating mathematically an individual's peak level of productivity and its inevitable decline before the person dies carried an unprecedented fascination.

"Marchetti claims to have investigated close to one hundred individuals and found their productivity to proceed along S-curves," said Ted. "I was able to obtain confirmation for over a dozen men from the arts and the sciences, people such as Bach, Schubert, Brahms, and Einstein. But one of the most fascinating cases was Mozart. Apparently, when Mozart died at thirty-five years of age, he had already said what he had to say.

"The data I considered were all of Mozart's compositions carrying a Koechel number. The fit turned out to be good. I found an S-curve that passed impressively close to all thirty-one yearly points representing the cumulative number of compositions. There were two little irregularities, however; one on each end."

Ted showed them a graph that he had brought with him.

"The irregularity at the low end of the curve required the inclusion of an early-missing-data parameter. The reason: better agreement between the curve and the data if eighteen compositions were assumed to be missing during Mozart's earliest years. His first recorded composition was created in 1762, when he was six. However, the curve extrapolated to its nominal beginning, i.e. the 1 percent of the maximum, at about 1756, Mozart's birth date. The conclusion is that Mozart was composing from the moment he was born, but his first eighteen compositions were never recorded due to the fact that he could neither write nor speak well enough to dictate them to his father.

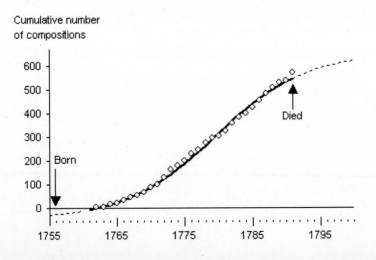

"The second irregularity was at the high end of the curve. The year of his death 1791 showed a large increase in Mozart's productivity. In fact, the data point was well above the curve, corresponding more to the productivity projected for the year 1793. What did Mozart try to do at the end of his life? His creative potential had been determined as 644 compositions and with his last work his creativity would be 91 percent exhausted. Most people who die of old age have realized 90 percent of their creative potential. There was very little left for Mozart to do. His work in this world had been practically accomplished. The irregularity at the high end of his creativity curve indicated a sprint at the finish! What he had left to do was not enough to help him fight the illness that was consuming him. We could say that in a way Mozart died of old age!

"In discussions with musicians, I have found that many are not shocked by the idea that Mozart may have exhausted his creative potential at the age of thirty-five. He had already contributed so much in every musical form of the time that he probably could have added little more in another fifty years of calendar time. He himself wrote at the age of twenty-one: 'To live until one can no longer contribute anything new to music.'"

A little later Ted brought back one of the topics they had discussed in one of their earlier Mendios meetings: Christ's whereabouts between the age of fifteen and thirty. He showed them that the evolution of the canonization of the catholic saints was also amenable to an S-curve analysis. The process consisted of two large cycles of *natural growth* spanning many centuries each. But the first one, known as the patristic ecclesiastical wave, had its roots extending before Christ all the way back to 300 B.C.

"Here we have evidence that Christianity began before Christ," concluded Ted. "The rate of growth in the number of saints during the first few centuries A.D. seems much faster than the *natural* rate observed later on. One may interpret this as a 'catching-up' effect. People had been expecting a savior for a long time. When one finally came, a certain amount of pent-up energy was released, resulting in an accelerated rate of sainthood attribution during the first few centuries AD. Such accelerated growth is often observed at the beginning of an S-curve. The starting up of natural growth might be hampered for a variety of 'technical' reasons. But once the growth process is underway, it proceeds faster than normal for a while to make up for the time lost.

"Had canonization proceeded all along according to the more natural rate observed later, it should have its beginning *before* the birth of Christ. If the patristic curve is backcasted, its nominal beginning (the 1 percent of the maximum) falls somewhere around the 3rd century B.C.

"According to Marchetti's explanation, the Judaic schism of the Essenes was active at the time in that region and, as individuals or in brotherhoods, demonstrated asceticism and extraordinary piety from sometime during the third century B.C. until the first century A.D. Information from the Dead Sea scrolls indicates that their doctrine contained many of the essential elements found later in the Christian message. It is possible that Jesus matured among them during his fifteen-year absence."

Aris and Mihali listened insatiably. Aris in particular was enchanted. His old theory about Christ having drawn secret knowledge from Ancient Mysteries was now scientifically, if only partially, vindicated.

28 – Symmetries and Invariants

Ted continued discussing all the invariants Marchetti had bombarded him with during their meeting in Vienna. Invariants are like conservation principles. Something that remains constant independently of time and place is as if it were conserved. But behind every conservation principle in physics there is a fundamental symmetry. The well-known conservation of energy is a consequence of the symmetry in time, namely the fact that the laws of physics do not depend on *when* we study them and would continue to be valid even if time ran backward. Another principle, conservation of momentum, responsible among other things for the clean-cut way billiard balls transfer their speed upon collision, is a consequence of the symmetry in space—i.e. space's homogeneity—in other words, the fact that things happen equally well in all three space dimensions. The two variables, the one conserved and the one depicting symmetry are called *conjugate* variables. Energy is conjugate to time, and momentum is conjugate to space.

But what symmetry would be responsible for conserving something like the number of deaths from car accidents? Marchetti had pointed out that this number grew with the appearance of cars until the mid 1920s, but from then onward the annual number of deaths had been confined to not much more than twenty per 100,000 inhabitants in most countries around the world. A homeostatic mechanism seems to have entered into action when this limit was reached, resulting in an oscillating pattern around the equilibrium position. The peaks may have produced public outcries for safety, while the valleys could have contributed to the relaxation of speed limits and safety regulations. It is remarkable how persistent self-regulation on car safety has been in view of tremendous variations in car numbers and performance, speed limits, safety technology, driving legislation, and education.

Car accidents are weighted against car benefits, and the final level of accidents is the price society is willing to pay for the use of cars. This price—like all prices—is by definition right, that is, society would not be happy with fewer accidents if that meant less car use. After all, deaths from car accidents could be eliminated altogether, if all cars where to be abolished.

But what symmetry could be responsible for the "conservation" of car accidents and what variable is *conjugate* to car accidents?

Ted thought for a long time until he arrived at the following conclusions.

Car accidents are conserved—i.e., they remain invariant over time and place—because there is a symmetry in the way cars are used, namely, *everyone everywhere* can own a car *anytime*. Good or bad drivers, young or old ones, all can drive a car. Had cars been the sole privilege of, say, the armed forces or the politicians, the annual death rate would not be stable over time but rather go up and down with wars or elections, as the case may be. Car accidents are conjugate to the right to own a car.

Similarly, the invariants concerning the daily quota for traveling have to do with human well-being. An expenditure of 15 percent of one's income is equivalent to an expenditure of 15 percent of one's energy. Expenditure of energy in daily displacement is conjugate to the homogeneity of human beings. The energy expenditure is conserved because there is a "symmetry" among people: they all generally sleep 8 hours a day, work 8 hours, and spend 8 hours at home, independently of race, nationality, or location.

Finally daily travel time is conjugate to daily expenditure of energy. Travel time is conserved (at around 70 minutes a day) because of another "symmetry" among humans: they all spend 15 percent of their energy in daily traveling. In fact, the two figures are related if we consider that *movement* takes place mostly during working hours and 15 percent of 8 hours is 72 minutes.

Ever since his discussion with Marchetti, Ted had become sensitive to the possibility of witnessing conservation principles in social situations. For example, one's happiness is often linked to someone else's misery. A joke is generally made at someone's expense. In comedy people are entertained and laugh at others' misfortune. Not unrelated is the fact that those who become rich generally do so on the expense of those who become poor (conservation of wealth). But just as both rich and poor belong to a healthy economy, both happy and unhappy seem to belong to a harmonious society, as if some amount of collective suffering cannot be avoided because of a deeply rooted principle of conservation.

It is not possible to establish a principle of conservation in a social context as rigorously as in physics. The difficulty with ascertaining a social absolute truth is the subjective element inherent in human judgment, the variation from one individual to the other. Yet, observations repeated over the years increase objectivity. The ultimate distillation of such knowledge becomes cast in proverbs that are endowed with unquestionable validity, and enjoy a status comparable to those of laws in physics. Guided by common sense and popular wisdom, Ted was able to identify a number of social conservation principles. Some of his observations were naïve; others more sophisticated. In each case he tried to understand what symmetries lay behind and might be responsible for the phenomenon. He searched for the "coordinate" that would be conjugate to the "coordinate" observed to be conserved; finding it could serve as a key to unlock secrets about what is really going on.

One of his early observations was on his young children while fighting and crying. Whenever his son hit his sister, she would cry. Now if Ted punished him, she would not stop crying until his punishment became severe enough for him to begin crying. In turn, an efficient way to make him stop crying would be to go back and scold her until she cried again. So crying would move back and forth from one child to the other. One stopped when the other started. They would neither cry together nor stop crying altogether. Ted called it the conservation-of-crying principle among young siblings. He was able to trace it to the "symmetry" in the fact that he loved them both equally. Children's crying is conjugate to impartial parental love. Had he loved only one of them, the other albeit unhappy wouldn't be crying continuously for not knowing better, and Ted would have never observed a conservation of crying.

A very similar situation occurred with adults. One day Ted was frustrated with Gilda, the group's secretary at work, because she told him that she was too busy and that he should make his own travel arrangements. He raised his voice and argued that his salary was bigger than hers exactly because he was expected to spend his time in other ways than making travel arrangements. She felt insulted and rushed to her desk crying. It wasn't long before Walter, their boss, was brought in. He summoned both of them into his office and demanded that Ted apologize to Gilda.

"I can afford to be generous," responded Ted arrogantly, "I'll apologize all you want."

This made neither Gilda nor Walter happy. "No," argued Walter, "you must apologize seriously."

A second light-hearted response from Ted made Walter angry. He persisted reprimanding Ted until the latter felt bad and only then he accepted the apology. Gilda was also now visibly satisfied. Ted's feeling bad had been a prerequisite for Gilda to stop feeling bad. Here too the conservation of unhappiness reflected the symmetry in the boss's desire to care *equally* for all his subordinates irrespectively of their rank. Employees' rivalry is conjugate to the supervisor's evenhandedness.

In society in general unhappiness is conjugate to human rights. Conservation of social misery and the fact that for some to be happy others must be unhappy is a consequence of the "symmetry" in everyone having equal rights. As if it were some kind of material, happiness cannot be spread thin and distributed to everyone. It can be accumulated somewhere if and only if it is depleted elsewhere. After all, "happy" really exists only when compared to "unhappy." In contrast, in a repressive society conservation of misery in the sense that "if you get it, I will not have it any more" cannot be observed because there are no exchanges in the collective happiness. The privileged remain privileged and the unprivileged remain unprivileged forever. They both get used to it and to a first approximation there are no manifestations of contentment or discontent on either side.

On a larger scale, society's legal system is also characterized by some principle of conservation of happiness. The law makes sure that undeserved acquisition of pleasure, such as by stealing and raping, becomes "compensated for" with imprisonment. The larger the offense, the more severe the punishment. The symmetry behind this conservation is society's aspiration to look after all its citizens *equally*. Citizens' well-being is conjugate to the fairness of the legal system.

On an even larger scale, one can argue that the often-cited religious concept of after-life reward or punishment also constitutes a conservation-of-happiness principle. In the Christian religion hard-working ascetic life leads to heaven whereas pleasure-seeking debauch leads to hell. But even the more elaborate eastern concept of reincarnation has bad reincarnations interlaced with good ones, as if meant to compensate for one another. Most religions feature a Devine justice of some type via conservation of ultimate happiness transcending death. The symmetry behind it in all cases is God's *equal* love for all people. Human happiness is conjugate to an omni-loving God.

Wealth is conjugate to free enterprise. Conservation of wealth in the sense that there are rich people and poor people and wealth will go from one person to another is a consequence of the "symmetry" in everyone's right to become rich. If this right were revoked, for example, in an ideal communist society, there would be no differentiation between rich and poor and no indication that wealth is conserved.

Concerning the inevitable presence of war in human society it must be said that war's justification has at times been most inappropriate. In fact, one frequent reason for war has been peace! There have also been wars for love, for Christ, for the Holy Land, and so on. Anything could become a valid reason for war, which raises questions of causality. Do wars have real causes or do they represent a fundamental necessity of life and can thus be triggered by *anything?* Is there a conservation-of-war principle implying that wars must always come and go and there can be no way to eliminate them once and for all?

Wars are conjugate to cultural identity. That is, the symmetry that characterizes the members of a cultural group—the common collective programming of the mind—gives rise to the existence of war between groups. In the kingdom of animals same-species groups do not generally wage war on each other. The socio-biological equivalent of "species" is "culture." In a society made up of only few representatives from all cultures, no warlike actions would be expected other than squabbles between individuals. War occurs between different cultures. The conclusion is discouraging. War will disappear from the face of the Earth only when cultural differences disappear. The price for the desirable cultural diversity seems to be the inevitability of war.

In manufacturing, product quality is conjugate to workers' motivation. That is, the quality of products produced by a certain class of workers will be "conserved," i.e., will be the same from one industry to another, or from one country to another, as a consequence of a symmetry—common feature—in the group of workers. Some people work to live. Others live to work. Still others work to make the time pass, for example, prisoners or mental patients. There is a product quality associated with each group, and it will be the same no matter what, where, or when they are manufacturing it. The motivation level of workers becomes the "price" of quality for the products they produce. A similar explanation may hold for the argument that one's genius would have also shown up in fields different from the one he or she became known for; a principle of conservation of genius. It suggests a possible conservation of creativity that shows up no matter where the person may be. If one spends prime time and energy creating a piece of art, prose, or music—let us say instead of creating a family and raising children or instead of performing heroic acts at war—the quality and creative content of the final product is likely to be high. If, on the other hand, the artistic expression is a result of a hobby, having nothing else to do, or simply boredom, the final result is likely to bore its audience.

Biodiversity is conjugate to competition. The symmetry here is in that survival of the fittest is equally valid for all species; there is no favoritism. As a consequence, biodiversity is naturally conserved. An exception is *Homo sapiens*, who, thanks to his intellect, has become less and less subject to natural selection over time (for example, many "unfit" babies survive nowadays). As a consequence, the conservation of biodiversity becomes eroded. The price of conservation in this case is our capability to violate the law of natural selection. And since we are not likely to give up this capability voluntarily, the statement should be turned around: the price of humans rising well above other species constitutes a reduction in biodiversity.

When Mihali heard the above discussion about conservations and symmetries he suggested the next logical step, namely, to look for and interpret the "action," which is defined as the product of the two conjugate coordinates in each case. Typical actions are the products (momentum) x (space) and (energy) x (time). The former is intuitively obvious: the faster one moves and the further one gets, the more actions he or she achieves.

But also the action behind the product (wealth) x (free enterprise) makes intuitive sense; a characteristic picture is the action-filled trading floor at the stock market. Action is also evident in the product (war) x (cultural identity); the phrase "missing in action" seems appropriate here. In fact, the formula indicates that a large war between similar cultures (little differentiation in cultural identities), like World War I, may result in a comparable amount of action to a smaller war between very different cultures, like the Vietnam War. Indeed, if we take the US casualties as a proxy for action, the two wars differ by less than a factor of 2 (110,000 casualties in World War I and 58,000 in the Vietnam War).

However the action expressed by the product (biodiversity) x (competition) seems rather intriguing because action implies a subject, an organism, or other entity behind it. Who could this entity be? And then, action's rate of change represents energy. Trying to comprehend action and energy in this example requires plenty of imagination!

29 – The Shape of Things

The excited discussion of Ted, Aris, and Mihali made them lose track of time. When Ted looked at his watch it was twenty-five minutes before his airplane was scheduled to take off. He rushed into a taxi and offered the driver incentives for pushing his new Mercedes to the limit in order to get to the airport terminal as fast as possible.

"It is not possible to go faster," complained the taxi driver. "If it isn't a traffic light, it is a traffic jam. One way or another you can't go fast in urban areas. Hundreds of horsepower under the hood accelerate you only to decelerate again a few yards down the road. I bet you a model A Ford would have gotten you there in the same time," he kept grumbling.

They barely made it to the airport for Ted to catch his flight. As he finally relaxed in his seat in the airplane he could not get his mind off the concept of invariants. The taxi driver's grumbling seemed justified. But Ted discovered only later that indeed the average urban car speed in the United States is about thirty miles per hour and has hardly changed since Henry Ford's time. Another invariant! Whenever faster cars or new roads are built, the number of circulating cars increases until the average car speed drops to thirty miles an hour. It then becomes unattractive for additional cars to get on the road and people try to arrange their lives differently. When that becomes too difficult, pressure builds up for more roads to be constructed.

What could be the symmetry behind this invariant? Perhaps the homogeneity among cars, the fact that they all take up about the same space on the road and they all have about the same performance. The average car speed is probably *conjugate* to the number of cars on the road, but whereas the former remained invariant since Ford's time, the latter increased every time new roads were built and/or new car technologies employed. As for action, the product (average car speed) x (number of cars on the road) seemed to have plenty in it.

Ted realized that invariants often represent the ceiling of an S-curve, the equilibrium between opposing forces. Rabbits will multiply to populate a fenced-off meadow but when they become too numerous, competition for a limited amount of grass slows down the population growth via increased kit mortality, diseases, and lethal fights between overcrowded rabbits. From then onward an auto-regulating mechanism maintains the number of rabbits around the capacity of the ecological niche, a population *invariant* over time.

Such cases of equilibrium abound in society. Deaths from car accidents grew along an S-shaped curve to reach their invariant of an annual of about 20 per 100,000 inhabitants. The ceiling of that S-curve was reached in the late 1920s and from then onward accidents were auto-regulated, remaining rather constant while the number of cars kept growing.[*]

Auto-regulated equilibrium, known as *homeostasis* in biology, is effectively an invariant. It is the simplest thing to forecast because it doesn't change with time.

The more Ted thought about invariants and reaching a final equilibrium via an S-curve the more he realized that this process might be at the heart of the mechanism that shapes the form of most things around us.

The average car speed of thirty miles an hour can be combined with another invariant, the seventy minutes available for daily traveling, to yield the size of our cities. The average distance covered per day is the natural limiting factor in defining the size of urban areas. Communities grow around their transport systems. If it takes more than seventy minutes to get from one point to another, the two points should not reasonably belong to the same community. Cars permitted towns to expand. When people only traveled on foot, at three miles per hour, towns consisted of villages not much larger than three miles in diameter.

There was a factor of ten in speed between foot and car transport, and cities such Los Angeles attest to the 30-miles-in-diameter size. Airplanes expanded the limits of urban areas further, and it is possible today to work in one city and live in another. Air shuttle services have effectively transformed pairs or groups of cities in the United States, e.g., Boston and New York, into one large "town."

It is not only the size of "towns" that reflects the existence of equilibriums encoded in invariants. The final shape of most manmade objects results from some kind of equilibrium of opposing forces, for example, electronic pocket calculators and ballpoint pens. As technology advanced, calculators became smaller and thinner; ballpoint pens finer. They finally reached stable dimensions but only after experimenting with extreme ones. At some point ultra-thin calculators came out—as thin as a credit card—and super-fine ballpoint pens that drew lines a few thousandths of an inch thick. They were technological marvels and yet, a few years later, they were no longer available in the

---

[*] Toward the end of the 20th century deaths from car accidents declined significantly but that was due to the fact that air travel signiciantly replaced tarveling by car.

mass market. They turned out to be unsuccessful products. They were too extreme, not for the technology, or the manufacturer, but for the user. The calculators proved too fragile in people's wallets, and the ballpoint pens pierced and scratched the paper. The low demand could not justify continuation of their mass production. The opposing forces were balanced at the final sizes, later settled at less extreme values: quarter-of-an-inch thickness for the calculator, and the size dubbed "fine" for the ballpoint pens.

Similarly, the forms and shapes of cars, roads, sidewalks, and houses evolved through many generations, and slowly settled at their final configuration. Knowledge accumulated over the repetitions and under the exertion of forces reflecting climate, culture, pedestrian demand, traffic needs, standard of living, etc. In the process, they invariably overshot their final equilibrium position but then fell back and oscillated as they "hunted" for the optimum shape. In a way, the intermediate states served the purpose of mutations that became selected out so that only the best-fit one survived.

The emergence of an optimized final shape requires repetitions. Doing something over and over again (with feedback) converges on a final shape (via competitive selection among intermediate shapes). It is the same process as evolutionary selection in biology. Moles do not have eyes because they do not need them in the dark underground. Proverbial wisdom has encoded it in the dictum: Use it or lose it.

Every environment molds the species it supports. Fish become bigger in larger aquaria. Cars in America are bigger than in Europe because roads are bigger. (Marchetti has demonstrated that contrary to popular belief roads preceded cars in their diffusion process.) The most successful virus is not necessarily the most or the least virulent. It's the one that exploits the host most effectively. Evolutionary biologist Paul Ewald has argued that a Darwinian perspective can enrich our thinking about the evolution of the HIV virus. Condoms could conceivably push the virus toward more benevolent forms, simply by depriving virulent strains of the high transmission rates they need to survive.

Homeostasis involves a random search in extreme directions. Equilibrium among opposing forces necessitates corrections after having gone too far. As a consequence there are fluctuations that throw a chaotic vale over the invariant auto-regulated level. Still, despite the chaos there is predictability and the reassuring element of harmony and well being behind *homeostasis*.

While the easiest thing to forecast may be the level of an equilibrium position, the opposite is also true. One can also easily forecast something that is moving. Motion enjoys its own stability; the phenomenon is called *rheostasis*.

This is obvious when it comes to a bicycle that cannot stand up unless it moves. All motion implies stability. Businesspersons often use the super-tanker metaphor. It is easy to predict where a supertanker will end up in the near future because it takes so long for it to change direction. And there are more "exotic" situations. Physicists characterize high-energy particles as "stiff" because it is difficult to bend their trajectory. In fact, stability is not related simply to the speed but to the *momentum*, that is, the product of speed times the mass. Supertankers go relatively slowly but are very massive. Elementary particles have miniscule masses but travel at dazzling speeds. They both enjoy stability in their motion.

For a natural growth-process things also become predictable during the fast-growth phase around the middle of the S-curve pattern. A business venture well under way and going at full speed is hard to modify or stop and leaves management with restricted freedom of choice.

"To be the head of a large business is like having mounted a lion. You simply have to go where it wants," a chief executive officer once confided to Ted. But Ted understood more than what the CEO affirmed. If that CEO had little choice, it meant that his company was enjoying rapid growth—i.e. was around the middle of its S-curve—otherwise there would be plenty of choice for action.

30 – Natural Growth in Competition

Less than a year after Ted abandoned the scientific environment of CERN he was feeling already completely integrated in his new setting. He did not particularly mind wearing a jacket and a tie every day, and he arranged it so that he could stay at work late in the evenings or go to his office on weekends whenever he wanted to. He was making a niche for himself both in what concerned the style and the content of his job. He even began appreciating what he was learning about the business world. But his favorite subject matter was using S-curves and other unconventional techniques to forecast the future.

An S-curve and the associated life cycle are two different ways of looking at the same growth process. The S-curve represents the size of the growth and points out (anticipates) the growth potential, the level of the final ceiling, how much could one expect to accomplish. The bell-shaped life-cycle curve represents the *rate* of growth and is more helpful when it comes to appreciating the growth phase you are traversing, and how far you are from the end. The S-shaped curve reminds us of the fact that competitive growth is capped. The bell-shaped curve reminds us that whatever gets born eventually dies. From an intuitive point of view, an S-curve promises a certain amount of growth that can be accomplished, whereas a bell-curve heralds the coming end of the process as a whole.

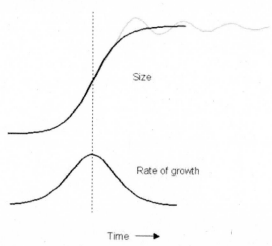

As for invariants, they indicate the equilibrium of opposing forces, and they manifest themselves at the ceiling of the S-curve after the growth phase has been completed; they also reflect optimization and a state of well being.

Little by little Ted developed an intuitive appreciation for these patterns and began detecting their presence in many everyday situations. During one of the Friday-afternoon get-togethers of the DEC employees in the small building, where Ted had his office, there was much talk about new desks and office equipment; it culminated as follows:

"We all know how painful it has been to get our coffee from a machine that requires exact change. How many times has each one of us run around asking colleagues for coins? Well, you can consider this activity part of the past. Next week we are installing new coffee machines that accept any combination of coins and provide change. This should improve employee productivity and satisfaction."

In spite of its banality, the statement resonated in Ted because he had indeed been frustrated more than once by not having the correct change for his mid-morning cup of coffee. The announced improvement had been long overdue. The following week the new machines were installed and for several days employees made jokes about mechanical intelligence and technological comforts.

It was several weeks later that Ted found himself in front of the machine without *enough* change for his coffee. He looked in his desk drawers for any odd change—in earlier times there would have been "emergency" coins lying around—but with the new machines, there was no longer any reason to hoard small change. Ted's colleagues were in a meeting and he did not want to interrupt them with his petty problem. He walked over to the secretary's desk; she was not there. Instinctively, he looked in the box where she used to keep change for coffee; it was empty.

He ran out of the building to make change at a nearby restaurant, wondering why those coffee machines couldn't accept paper bills. Then he realized that the old machine had not been so inconvenient after all. It had been a nuisance to have to plan for a supply of correct change, but since that was the reality, they all made provisions for it in some way or another. It took some forethought, but one could not really say it was a stressful situation (being deprived of a cup of coffee never killed anybody). With the installation of the new machines, however, they relaxed more. Perhaps Ted had relaxed a little too much. He decided to make sure that in the future there was always some change in his desk so as to avoid situations like this.

At the same time he felt there was a certain quota of inconvenience associated with getting a cup of coffee. If the amount of inconvenience rose above the tolerance threshold, actions were taken to improve the situation.

On the other hand, if obtaining a cup of coffee became technically too easy, people would relax more and more until problems crept in to create the "necessary" amount of inconvenience. Imagine a situation where coffee is free and continuously available. A busy person, trying to be efficient and exploit the fact that there is no longer reason to "worry" about coffee, may simply end up not planning enough time to drink it! Perhaps people are not happy with too little inconvenience. Conservation of nuisance! Ted smiled at the thought. What symmetry could lie behind a conservation-of-nuisance principle in getting one's coffee? It must have something to do with the fact that the same amount of attention is solicited in a competitive way by all other activities of importance comparable to getting a cup of coffee.

It is true that many people spend all the money they make in order to cover their needs. Yet this becomes the case again soon after they get a sizable salary increase. The usual rationalization is that their needs also grow so that they still spend their entire income. The fact, however, that they always spend just what they earn, no more, no less, points to an equilibrium in which the needs grow only to the extent that there is more money. The same may be true of work, which expands to fill the time available. Some people, for example, continuously find things that need to be done around the home, independently of whether they went to work that day or not.

A month's salary can be seen as a "niche" to be emptied via expenditures. A day's free hours can be seen as a "niche" to be filled by errands and work. But no niche in nature will remain partially full or partially empty under *natural* conditions. Natural-growth processes proceed to completion along S-curves. This is why S-curves, bell-shaped curves, and invariants all possess predictive power. Becoming familiar with them, in an intuitive way, brings out the continuity in many fundamental processes, and thus eliminates a certain amount of uncertainty and anxiety about the future. It is reassuring, for example, to know that car accidents, just like AIDS victims, are auto-confined to a level that is not simply tolerable but also *desirable* to society. It is clear that society could reduce them further, but there are other priorities, and, in a way, society "prefers" these levels to remain where they are.

But it is also obvious that the path to the equilibrium level must be a smooth and continuous trajectory along an S-curve; it is not possible to instantly jump to the final level of the ceiling. Deviations from the natural path should expect to meet resistance often coming from surprising directions. Ted had found several examples of natural-growth processes that had been tampered with and had resulted in confrontation.

Nuclear energy entered the world market as a primary energy source in the mid 1970s when it reached more than 1 percent share. It grew along an

S-curve like all previous sources (wood, coal, oil, and gas) but its rate of growth was disproportionately rapid, compared to the other four all of which conformed closely to the same more gradual rate. Nuclear energy's high rate of growth was probably due to the craze triggered by the demonstration of nuclear power with the use of the atomic bombs in the end of World War II. The high rate was also probably responsible for a number of serious nuclear-plant accidents that in turn triggered vehement opposition from the environmentalists. As a consequence of the intense criticism, the growth of nuclear energy slowed down but did not stop. Fifty years later nuclear energy faces less vehement an opposition from the environmentalists who became more concerned about CO_2 emissions.

In the same energy picture coal declined as a primary energy source since the turn of the twentieth century. In the United Kingdom a downward point-ing S-curve was well-established for the production of coal by the mid-1970s. It suggested that production should continue dropping to reach less than 20 million tons a year by the end of the twentieth century. For the UK govern-ment, however, such a vision was completely unacceptable and in 1975 it halted the *natural* decline in coal mining by a legislative act that fixed produc-tion at 125 million tons a year. The act caused a clear deviation from the declining S-shaped trajectory of coal production, which lasted nine years. But at the end miners staged the longest strike ever, bringing coal production down to what it should have been had it followed its S-shaped pre-decree course. Despite renewed high-production levels after the long strike, coal production in the United Kingdom by the end of the twentieth century was very close to the 20 million tons a year, forecasted thirty years earlier.

In the light of the difficulties and the resistance encountered whenever natural-growth processes are not respected, it is not surprising that attempts to abruptly install democracies in countries such as Afghanistan and Iraq have met with violent opposition and ended up unsuccessful. It is a conse-quence of trying to jump to a final state without following the smooth evolution of the S-shaped trajectory of natural growth.

31 – Showing Off at Conferences

It wasn't long before Ted began attending conferences and presenting papers again. The conference topics now were no longer particle physics but Forecasting and Operation Research. Yet the style of the work was similar. These gatherings were generally frequented by academic people, and Ted took particular pleasure in provoking them with his new pragmatic point of view of someone coming from industry. The fact that he worked for a very successful computer manufacturer facilitated his acceptance as a speaker and the fact that he had a Ph.D. in physics propped him up during his audacious excursion off the beaten track of the disciplines he was now treading on. Physics and science credentials provide respect that can be used as counterweight in making bold assertions.

Geneva's association of medical doctors invited him to one of their yearly forums. The event organizer asked Ted to talk about predictability in the diffusion of diseases and the effectiveness of vaccines. Ted chose his talk's title as, "Can There Be Health without Medicine?" He talked about how S-curves and invariants illuminate the way diseases come and go. He showed evidence that vaccines are generally not discovered before the disease is well on its declining course. He was delighted to see how quickly the doctors in the audience caught on to his ideas. The speaker that followed him at the podium brought up among other things the topic of in-hospital infections that constitute a nuisance for the medical world.

"It could be that the reason we seem unable to reduce in-hospital infections much below the level of 10 percent is what we just heard, namely that this number is an invariant and represents some kind of homeostasis," he said.

Another conservation principle, Ted thought, and his mind began working. What could be the symmetry behind it? And what would be *conjugate* to in-hospital infections?

The answer was not difficult in this case. In-hospital infections are largely due to the fact that patients in hospitals become exposed to new viruses and germs other than the ones they are carrying. In-hospital infections must be conjugate to hospital's ability to handle many different diseases. The symmetry behind this conservation principle is that one can go to a hospital for *every* type of disease. If an institution treated patients with only cirrhosis of the liver, or only tuberculosis, it would have a negligible rate of in-house infections.

During coffee break several doctors came over to Ted with appreciative comments about his presentation. The positive remarks delighted him. With his ego pumped up, he noticed two attractive young women who seemed almost out of place in this gathering heavily dominated by men. With a coffee cup in his hands he worked himself through the crowd toward them and asked the obvious question.

"We are plastic surgeons," was the response that came as if to serve more than just answering his question.

Ted was triggered to dish out another one of Marchetti's provocative invariants even though he did not really believe in Marchetti's far-fetched explanation. "Empires broke up in two when they grew so big that it took more than two weeks for a messenger to reach the outposts. This is what happened to the Roman Empire. Even though Romans could reach the Black Sea by boat in two weeks, pirates forced them to travel by land and that took about a month. That's why Byzantium was created."

"Why two weeks?" asked one of the women. "What is special about this period of time?"

"Because the round trip would take 28 days and that corresponds to women's menstrual cycle. Anthropologically speaking males have to reinstate their dominance every month at the right time of the month. That is how hierarchies have been maintained. Today's practice of monthly CEO meetings could be a vestige of that phenomenon."

Ted's interaction with the French-speaking medical world did not end there. After the forum in Geneva his presentation was written up, translated into French, and published in the medical journal *Médicine et Hygiène*. A few months later he received a letter from the president of the French National League for Liberty in Vaccination, an organization devoted to fighting the institution of obligatory vaccinations. They wanted permission to reproduce Ted's article in their journal and to attach it in a memorandum they would submit to the French Parliament. Ted gave his permission feeling that he was making history.

The same year he attended another conference at the World Health Organization with the theme "Consultation on Health Futures." A number of speakers took turns at presenting scenarios for alternative futures concerning health topics. When Ted's turn came he began his talk as follows:

"It puzzles me that everyone here refers to the future in the plural as 'futures' and I find it disconcerting that this word has entered even the theme of our conference. There has been only one past and I guarantee to you that there will be only one future. Our task as forecasters is to anticipate it as well as possible. Otherwise we are indulging in academic exercises and I have learned painfully that in industry academic is a synonym to useless."

He then presented S-curve forecasts on a number of issues with estimates for the uncertainties but no scenarios or alternatives.

His talk was received unevenly. Many in the audience nodded in approval as Ted talked. Some of them came up to him later to express their appreciation. But the majority—almost all academics—tried to brush Ted aside as a maverick.

That evening during the conference's gala dinner and in a more informal social setting Ted amused himself further by going against his own words.

"No one corrected me this morning when I said that there was only one past. Of course it is not true! The past keeps changing all the time."

"What do you mean, how can the past ever change?" wondered the wife of a quiet older Dutch professor at the other side of the table.

"Well, take Romania's Nicolae Ceausescu for example; he was considered to be a hero up to 1989 but after that date he was considered to be a tyrant," answered Ted.

"Hero and tyrant are our characterizations of him. He was what he was and that did not change," came back the lady. "Our opinion might change but history cannot change."

"I disagree," objected Ted. "History changes as soon as new data become available by accident, discovery, or new research techniques."

Now the lady raised her voice in frustration, "Discovering new data does not change what *really* happened. It only changes our knowledge of what happened."

"But that's all there is," insisted Ted. "Reality exists only as we know it. If a tree falls in the woods but no one sees it and there is no recording of the event anywhere, did the tree really fall?"

Then he remembered a science-fiction story that had made him ponder on this issue many years ago, so he added, "If there is a nuclear war and everyone dies except two persons and one person says it is daytime while the other one says it is nighttime, what is it in reality?"

Ted enjoyed the role of the provoker not the least because it helped him grab the audience's attention. He reveled in people listening to him. Conferences offered a good platform because besides the presentation he got a second chance with an audience at coffee breaks, dinners, and social events. But on one occasion he found his match. It was during a conference at IIASA where Marchetti had also presented provocative results. At dinnertime Ted sat at Marchetti's table where a theatrical Robert Vacca dominated the conversation. Vacca was another hard-core scientist, member of the Club of

Rome, and author of alarmist science fiction such as *The Coming Dark Age*. At one point Vacca asked Marchetti, "Cesare, when are you going to join the Club of Rome?"

"I cannot do that," replied an arrogant Marchetti, "it is a matter of self respect."

The two men continued exchanging "niceties" and even told dirty jokes. Ted wanted to join the conversation, and realizing that he was dealing with heavyweights he tried his best joke, the contest-winning one about the lion and the monkey that Aris had told in the sauna of Chandolin.

Marchetti and Vacca listened impassively; at the joke's punch line neither blinked an eye. Following an embarrassing moment of silence Vacca looked at Ted and said, "Go on!"

"What do you mean?" asked Ted puzzled.

"Well," continued Vacca, "the monkey episode got the lion very upset, as you may well imagine, so he began chasing the monkey around the jungle in order to eat him before he tells anyone else and the news spreads. During the hot pursuit the lion passed in front of an explorers' encampment where he saw someone lying in a hammock reading a newspaper.

"'Excuse me,' shouted the lion, 'have you by any chance seen a monkey running around here?'

"The man put down the paper, lowered his head to look over his glasses and asked, 'which monkey, the one that screwed the lion?'

"A demoralized lion responded, 'Is it already in the papers?'"

32 – Publishing for Fun, not for Survival

Besides conferences Ted considered other alternatives for publicizing his new hot subject. Despite the lack of enthusiasm for academic activities in his new industrial environment, he submitted two articles for publication to *Technological Forecasting & Social Change*, an international journal that Marchetti had recommended. One of the two articles included an extensive Monte-Carlo study on S-curves.

A Monte-Carlo study is a computer-intensive simulation program that allows for random events. Physicists use them to study the properties of their experimental detectors, such as precision and detection efficiency. Ted with a mathematician subordinate wrote such a program to determine the size of the uncertainties he should expect when fitting S-curves to data. He no longer had the mainframes of CERN at his disposal but they set several DEC minicomputers to run simultaneously for an entire weekend. On Monday morning they found piles of printout reporting on 40,000 S-curves fitted on artificial data purposely distorted by various amounts of "noise." It was the most extensive study on S-curves ever made.

The study yielded estimates for the uncertainties to be expected and also correlations between the steepness of the curve and its ceiling. Specifically, Ted found that to some extent it is possible to make an S-curve reach a higher ceiling by simply forcing its rate of growth to slow down a little. This was a significant result because many people were uneasy by the predetermination ingrained in S-curve models. They invariably wanted to know how they could make a curve reach a higher ceiling from the one predicted. Could Mozart have lived his life so as to accomplish a larger number of compositions, and if yes by how much? The answer Ted obtained through his Monte-Carlo study was quantitative, rigorous, but also commonsense: you can go a little farther by being careful with your resources but you cannot perform miracles.

There was much common sense in S-curves. At the heart of the growth-in-competition law lies the fact that the rate of growth is at all times proportional to the amount of growth already achieved *but also* to the amount of growth remaining to be achieved. That is why the life cycle is bell-shaped. Take the diffusion of an epidemic, for example. Whenever there is a new epidemic, the rate of appearance of new cases is small in the beginning because the number of contaminated people is still very small. Toward the end, when practically everyone is ill, the rate of appearance of new cases is again small because there are no more people left to be contaminated. Obviously the rate of appearance of new cases is maximal around the middle of the growth process when half the people are ill and the other half constitute potential victims. This bell-shaped life cycle is symmetric and consequently what goes up quickly comes down quickly, reminiscent of "easy come easy go."

Another commonsense aspect of S-curves is the fact that something must be already growing for its growth rate to be nonzero. It resonates with "you must have gold to make gold," or the more cynical "banks only lend to those who have money." It becomes more intricate when it comes to learning and the accumulation of knowledge. The rate of learning increases with the more knowledge you have; alternatively, you cannot learn anything if you know nothing, which argues effectively for the necessity of teachers.

It also argues in favor of another "law" Ted had discovered from his own experience: the more you do something the more you want to do it. This anti-intuitive message carries much truth. Surprisingly, one may feel sleepy after having slept much, just as one is less hungry at lunchtime if he or she skipped breakfast. A study on man's sex habits revealed that men masturbate more often when they have regular sex; abstention and monastic life are difficult only in the beginning.

Ted had become aware of this law quite young. When a professor in graduate school has asked a rhetorical question about fundamental truths, students volunteered such things as "work is good for you", and "things will always get worse" resounding the second law of thermodynamics namely that entropy will always increase. But Ted offered his "the more you do something the more you want to do it." Little he knew at that time how this would constitute the seed of his later life's central activity.

The second article Ted published outside hard-core science was entitled "The Normal, the Natural, and the Harmonic." It was inspired by the discussion that once had led Ted and Bjoern to conclude that to be normal is to be different. In this article Ted began by pointing out the great similarity between the natural life cycle and the well-known normal distribution (or Gaussian). But whereas the former reflects a natural law the latter is but a mathematical concoction. Karl Friedrich Gauss was a well-known man of science referred to as "the prince of mathematics" in nineteenth-century literature, and the bell-shaped curve he provided resembled very much the distributions with which people were concerned. In addition, it is obtained through an elegant mathematical derivation. For mostly historical reasons, people proceeded to put Gaussian labels on most of the variables on which

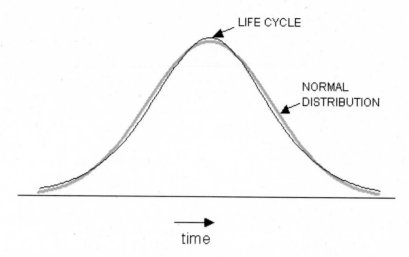

they could get their measuring sticks. In reality, however, there are not many phenomena that obey a Gaussian law precisely.

The natural life cycle is so close to a Gaussian curve that they could easily be interchanged. No psychologist, statistician, or even physicist would ever have come to a different conclusion had he or she replaced one for the other in a statistical analysis. What is ironic is that the Gaussian distribution is called "normal" while there is no natural law behind it. It should rather have been called "mathematical" and the name "normal" reserved for the S-curve law that is so fundamental in its principles.

This situation echoes something that happened in music. It was for convenience that Bach introduced the well-tempered way of tuning musical instruments. A mathematical genius, Bach was drawn to compose on all possible keys. The twelve semitones with the options of major or minor offer twenty-four different keys for composition. Up to that time fewer than a handful of them were being used by composers. The other keys produced dissonance for the normally harmonious chords of thirds and fifths.

Bach succeeded in finding a way of tuning his harpsichords that allowed him to compose in all twenty-four keys. The proof lies with his celebrated work *The Well Tempered Clavier*. The century that followed saw the widespread popularity of pianos as well as the appearance of brilliant mathematicians. Pianos made long-lasting tuning imperative, and mathematicians elegantly formulated Bach's way of tuning as setting the ratio of frequencies between successive semitones equal to $\sqrt[12]{2}$, (the twelfth root of 2). In so doing they produced *equal-temper* tuning, pushing Bach's idea one step further. (Contemporary musicologists argue that Bach may have disapproved of this!)

Thus transposing became possible, musical ensembles flexible, and life generally easier for all musicians.

As it happens, equal-temper tuning is very close to the natural octave, so that the differences pass by the ear almost imperceptibly. The natural octave, however, may have more profound reasons for existence. Purist musicians claim that piano and violin can no longer play together and sometimes demand an old-fashioned tuning for a sophisticated recital program. Furthermore, studying the natural octave may provide insights and understanding on a cosmological level. Gurdjieff presents ideas that relate the natural octave to the creation of the universe!

The seeker of pure and esoteric truth should be wary of elegant formulations such as those of Bach and Gauss.

Ted's two articles were published in the *Technological Forecasting & Social Change*. Ted continued submitting publications to this journal and it wasn't long before the journal's editor-in-chief invited him to join the journal's advisory board. Among his duties would be reviewing articles submitted for publication and occasionally writing reviews of new books. Ted was building a reputation as a forecaster scientist.

33 – Entelechy

At its zenith Digital Equipment Corporation enjoyed all the privileges of a successful company in an affluent society. The working force kept increasing and so did the remuneration and benefits. A swelling personnel department looked after the employees' well-being; in-house training became a priority. Each employee was encouraged to attend two to four week-long courses every year. It was considered as a form of additional and involuntary compensation. The content of these courses ranged from finance and accounting to time management and consultant skills.

Ted enjoyed attending courses because they invariably took place in fancy resort hotels in impressive settings, with good food and many new people. Occasionally he even picked up some new piece of knowledge, if of dubious utility. For example, in one course they were taught how to tell if their interlocutors are lying. You look at their eyes. If answering your question their gaze moves up and left, then they are telling the truth, but if their gaze moves up and right, they are making something up. Ted generally sent such pieces of information directly to his weirdoes cabinet.

One course about marketing was given in Fontainebleau near Paris, famous among other things for its culinary establishments. Besides good food there was a surprise for Ted. During one session of the course the speaker, a senior DEC employee, presented various forecasting techniques. Among others he presented S-curves and material from a DEC internal report Ted had written for the marketing department. When Ted saw his own graphs projected on the wall and clumsily explained he began interrupting the speaker with corrective remarks. A quick glance at the name card on Ted's desk was enough to make the speaker recognize an embarrassing situation. He promptly acknowledged Ted's contribution and invited him to continue the presentation of his own techniques.

The course was also attended by the person who was responsible for the organization of DEC's in-house training curriculum. He immediately decided to replace the disgraced presenter with Ted for all ensuing repetitions of the course.

Teaching a half-day module in that course proved for Ted a much richer experience than following the whole weeklong course. His session consisted of three hours of lectures in the afternoon and would always be followed by dinner with everyone in the evening. He amused himself and his audience.

What's more, following the first time, it required no additional work on his part. The event was repeated four times a year, each time at a different location in Europe.

Besides fun and games Ted realized that he was being drawn into the process of unveiling the future from different directions. As usually happens when one explores a new subject in depth, Ted became fascinated with the secrets he kept discovering. He was able to tie forecasting to the existential questions that haunted him, and even to Aristotle.

"Nothing happens without a cause," had been the teacher's opening statement in Ted's first biology class in high school. Ted retained pleasant memories from that course. The teacher was a brilliant biologist who talked quickly, made clever puns, was intolerant of slow responses, and turned every class into an adventure.

Ted came back to his teacher's opening statement affirming causality time and time again during that course but also later in his life. He remembers embarrassing his teacher once when he asked what is the cause that makes the heart pump. The teacher had to admit that it was rather complicated.

In the general understanding of causality the cause always precedes the effect. This sequence of events constitutes a sacred cow in physics, where violation of causality is tantamount to saying there is no science. And yet, Aristotle postulated that for many things the cause should be looked for in the future. Aristotle was not referring to rare unusual phenomena that negate science but to an unconventional form of "forecasting." Wings begin growing on baby birds in order to make flying possible *later*. During embryo development eyes begin to appear in order to make seeing possible *in the future*. "A wing is a very sophisticated 'forecast' of the dynamic properties of the air," wrote Marchetti, who considered forecasting to be the most important function of the nervous system. Many complex trajectories must be forecasted and thousands of muscles must be correctly timed by a hunting lion in order to succeed feeding itself *in the future*. A tennis player must accurately forecast where the ball will arrive *before* positioning his or her racket. Orangutans can make longer-range forecasts by memorizing where and when fruit will be ripe on the trees of their territory. All these forecasts make use of invariants that represent natural laws. Marchetti argued that DNA uses a model of the world, which gets adjusted *a posteriori* via natural selection. For example, men became bigger and stronger than women to facilitate hunting, which was the way of life for thousands of generations.

Obviously for these forecasts to be successful DNA's model of the world and its laws must not change rapidly over time. Physicists study similar laws to make their own forecasts, which they refer to as calculations.

Forecasting was proving for Ted to be an enormously rich subject with many facets. Not only all scientific theory revolved around the capability to predict, but also all living creatures developed according to what they will be doing in the future.

Aristotle's preoccupation with the way a final cause determines prior development culminated with his creation of a concept, which he called entelechy. The word contains the root of two Greek words "telos," which means "end" but also "purpose," and "echein," which means "to have."

The concept of entelechy brought back to Ted the old question of what the purpose of life may be. Aristotle seemed to suggest that the answer could be looked for in the very way life is made. Looking at how men and women are made, Ted's mother could have been right after all when she replied to her daughter's question with, "The purpose of life is to grow up and have children." Entelechy, S-curves, the ergodic theorem, and Ted's *Dum Possum Volo* law all shared a large conceptual overlap and all argued that the human capability to procreate would unavoidably result in procreation. Entelechy therefore decreed that procreation was one purpose of life.

But Gurdjieff argued that humans could do more than that. Ted could not forget Gurdjieff's apple-tree metaphor. An apple tree produces many apples, most of which will be eaten or rot away on the ground. In both cases these apples provide the useful function of nourishing other living organisms. But an apple contains the capability of giving rise to another apple tree, and despite the fact that it happens extremely rarely, entelechy argues that it is an apple's quintessential purpose.

Gurdjief used this metaphor to instruct his disciples to strive towards becoming more highly developed beings rather than food for others. As with apples, this could happen only extremely rarely, therefore the quintessential purpose of life couldn't be commonplace childbearing.

It had never been spelled out to Ted's satisfaction what these higher beings would be like. Nor had the entities to whom humans might serve as food or fertilizer been specified. But Ted found Gurdjieff's answer more attractive than his mother's even if the latter had Aristotle's blessing. Gurdjieff's was more ambitious, more exclusive, and displayed certain elegance like a mathematical theorem. Particularly so when Antonio, the Mexican Ted had met in his first camp at Chandolin, cast it for him in the form of an equation. Antonio was not a physicist but knew how to talk to one.

"The Work here consists of striving to achieve the state X," Antonio had said, "where X to being awake is as being awake is to being asleep," and he drew the following equation:

$$\frac{X}{\text{awake}} = \frac{\text{awake}}{\text{asleep}}$$

But Ted had placed all this now on the back burner. His main preoccupation had become forecasting and the arsenal of unique science-based methodologies he was accumulating for improving it. It was a topic of vital interest to a wide range of professionals, not the least businesspeople. He considered the possibility of teaching outside DEC in a business school. He liked teaching because, like conferences and publications, it gave him the opportunity to collect nourishing attention.

He made an appointment with Mike Anderson a social acquaintance who was professor of marketing at Geneva's International Management Institute (I.M.I.)

"I have some intriguing material to show you," he told Mike on the phone.

34 – One Cannot Part with Knowledge, Much Less with Understanding

It was an autumn afternoon. As Ted left the restaurant at which he had enjoyed a prolonged European-style lunch with Mike Anderson and other I.M.I. professors he told himself in an audible voice, "I am going to write a book!"

During the morning meeting at the I.M.I. campus they had explored the possibilities of Ted teaching there his avant-garde ways of thinking and his forecasting techniques. There had been much excitement in that meeting. Ted had talked about natural growth in competition, how it is manifested via S-curves that enter our lives in a thousand different ways, namely in every situation where there is competition. Products fill their market niche the same way rabbits fill their ecological niche. Creativity, productivity, and artistic achievement are all competitive growth processes and consequently they too follow S-curves.

"Did you know that Mozart died of old age?" Ted had asked them rhetorically. "When Mozart died, his creativity curve had practically reached its ceiling."

Over lunch everyone was in a witty and creative mood. "You work for a computer company and yet here you are talking to us about rabbits. What do computers and rabbits have in common?" asked the professor of business strategy.

As there was no answer he volunteered the punch line of his joke, "They can multiply!"

"You've got an S-cargo here," punned Anderson, pointedly eying his spicy hors-d'œuvre. "This is a sexy subject," he continued. "If you write a book and include a do-it-yourself diskette, it will become even sexier. You've got a bestseller here! Why do you want to waste your time teaching business executives?"

Ted was convinced. There over wine, desert, and wisecracks he decided that he was going to write a book that could well become a bestseller. The thought of tapping on public attention in unprecedented quantities was intoxicating.

In the weeks and months that followed Ted found himself devoting all his free time to book writing and even squeezing time out from his other obligations. His doctor's reaction was to the point when Ted told him that he was writing a book. "Oh, no, why do you want to do that? It is so antisocial. Ever since a friend of mine began writing a book he disappeared; we lost him!"

Ted's friends lost him too. Worse than that, he made an appointment with Michel de Salzmann to announce that he would stop going to the Gurdjieff group because he wanted to devote all his time to the book. He was apprehensive about this meeting because he suspected that Michel would once more present his argument on how one cannot break away from the Work.

He met Michel in the lounge of the Hotel de la Paix, sitting on the very same couch as they did when they first met more than ten years earlier. They talked about many of the things that had happened since then. Ted told Michel about his book project in an excited manner. He concluded by saying that he could no longer justify taking time to listen to the same people on the group saying the same old things.

"Do you think it is entertaining for me?" asked Michel. "We are not in the Work to search excitement. But a group of people is indispensable for our advancement. You cannot reproduce alone the conditions necessary for further personal development. Don't cut all ties with the group. You can join activities selectively, for example, only when I am here."

"But the energy and the emotions," argued Ted. "All these years I have been expecting to witness their emergence in some exercise or in the movements but instead I found them all while preparing this book. I have a feeling that I was looking for them in the wrong place." He stopped short of telling Michel that writing this book had become his purpose in life.

Michel felt Ted's exuberance and acknowledged his book project, "You have apparently found a bone. You should chew on it for a while." He paused and then continued more resolutely, "Stay in touch only with me. Write to me. Of course, I won't be able to answer all your letters."

There was something sad about saying goodbye to Michel like this. Ted vowed to remain in touch with him. But as soon as he left the hotel he rushed back to his desk and was immersed in the manuscript material. Michel and Gurdjieff were left far behind.

He often thought of writing to Michel and even composed letters in his mind and occasionally on paper, but never mailed any of them. Maybe that is what Michel had in mind in the first place; after all he said he would not be answering them.

Michel had also been right about not being able to break away from the Work. He had not said this explicitly to Ted, but the experience and knowledge Ted had acquired from his association with the Gurdjieff group was tantamount to having gone to a school, not unlike having studied physics. Schooling affects one's way of thinking for the rest of one's life.

Ted never stopped thinking as a physicist in his new work in the industrial environment, even in the home environment. He mapped scientific principles on everyday situations. For example, one day he had just finished shopping with his children and drove toward the exit of the underground parking lot when he realized that he had lost the magnetic ticket that would let him out. A considerable fine was at stake.

"What are you going to do?" asked worried his daughter from the back seat. "I'll increase my momentum," he answered cryptically, and then explained, "I will pick up speed and just sail through."

"You'll break the barrier?" more worried voices from the back.

He reassured everyone there was nothing to worry about, then drove up to the barrier and pushed the call button next the machine's microphone. When someone answered he said in a matter-of-fact voice: "Sir, I've misplaced my ticket. But I have paid it. It was fifty centimes. I've been here for a little over twenty minutes and I urgently need to bring my kids to the other side of town."

There was no answer but the barrier went up. The children were impressed with what their father referred to as "increasing momentum."

To be borne along by one's own momentum can be effective. Others cannot easily stop you or force you to change direction. You can go through barriers. Momentum is defined in physics as the product mass times velocity. The momentum can be large either because the mass is large or because the velocity is large or both.

Large momentum is crucial when you need to go past someone's assistant who is shielding his or her big boss. To increase your momentum you need to increase either your weight (mass) or/and your speed. The faster you move, or the heavier you are, the less resistance you will encounter. To bring this over to work environment let us consider the social equivalents for speed and weight. High speed corresponds to acting quickly. It often correlates to being bright. Weight corresponds to "heaviness" either because of high rank and clout, or because of a personal gift to impose your way. It is a characteristic of those people to whom you cannot easily say no. Big-posture individuals often get their way without necessarily having to be bright or to act quickly. In contrast, low-ranking managers, or people whose physique is such that you do not easily notice them, have to be sharp and move quickly in order to achieve their goals.

This becomes flagrant in presidential elections. A presidential candidate undergoes a fundamental change of posture once elected. Presidential candidates run around almost spastically as elections approach, but the one who is elected instantly becomes calm and slow moving. The "weight" of the new position allows him or her to be effective without having to worry about increasing "speed."

Ted had observed a correlation between high-level executives and their physical size. More often than not big bosses also were physically big. They were also slow-moving imposing personalities. In contrast, small-size leaders typically appear agile and hyperactive. But both weight and speed come into the momentum formula linearly. If you increase either one, you increase your momentum and your effectiveness by the same amount. One's "weight" may be more difficult to change, but when you find yourself physically exhausted from running continuously after people to make them do what you want, you may consider increasing your effectiveness simply by dressing up and behaving as if you were more important. You can increase either "speed" or "weight" depending on which one is more feasible and convenient.

A fascinating case in point, Ted had noticed, was physicist Carlo Rubbia, who besides being physically big, was also very smart, and talked very rapidly. His "weight" in physics was excessive and his momentum utterly effective. Despite an economic crisis, European governments were unable to refuse funds for his plans to build a giant accelerator at CERN.

Whenever Ted came across an everyday-life manifestation of a law of physics he tried to exploit his knowledge of the law to gain more insights and maximize the extraction of practical benefit. In the case of momentum, he knew that operating by time (i.e., dividing the change in momentum by the time during which this change takes place) yields an impulse, which translates to force. The shorter the time during which a change in momentum takes place, the greater the force. In everyday language, impulse means mental incitement, a sudden tendency to act without reflection, and has the notion of force behind it. The origin of this force can be traced back to a change of either the mass or the speed. A promotion is a sudden change in the importance of someone at work. It follows that the person promoted can suddenly exert more force on the people around him or her.

The world of business is full of slogans and practical advice. One is, "you want to abandon a sinking boat." But Ted knew that a sinking boat produces vortices that suck in nearby objects with a force, which depends on proximity. The closer you are, the harder you will be sucked in. The implication is that in abandoning whatever is sinking (boss, department, company, colleague, etc.) you need to introduce some distance between yourself and the sinking object. You should not be too close but at the same time you want to spare yourself unnecessary hardship and possibly missed opportunities by moving too far. How far should you go?

With the pulling force being inversely proportional to the square of the radius, you do not need to go very far. Increasing your distance by a factor of two will cut down the force—and consequently the risk—by a factor of four. Of course "distance" in a social context may be difficult to quantify, but in lack of better judgment, one may begin by taking the meaning of the word literally. If someone in your office is about to be fired, you can cut down your own risks by a factor of four simply by doubling your physical distance from that person.

35 – Setting Straight the AIDS Threat

Book writing became Ted's *raison d'être*. It was in his mind (and in his brief-case) wherever he went, be it work or vacation. He enjoyed weekends the most. His heightened sensitivity resulted in a snowball effect. New ideas would be triggered by random events. One day while driving and listening to the radio, he heard a discussion of Johan Strauss's *Die Fledermaus*. The piece was composed in 1873 but its debut was postponed for a year, and even then the public did not appreciate it. The reason was Vienna's stock-market crash in 1873, which cast a veil of economic gloom over society, at odds with the operetta's carefree content. It was only later that *Die Fledermaus* was recognized as a masterpiece.

The story grabbed Ted's attention because the date 1873 is exactly 56 years before the New York's stock market crashed in 1929. His Chapter 8 was devoted to this "cosmic" cyclical variation with stock-market crashes, energy prices, criminality, feminism, and many other human endeavors oscil-lating periodically with this frequency.

This 56-year cycle had first been mentioned in the literature by the Russian economist Nikolai D. Kondratieff, whose classic work in 1926 resulted in his name being associated with this phenomenon. From economic indicators alone Kondratieff deduced an economic cycle with period of about fifty years. His work was promptly challenged. Critics doubted both the existence of Kondratieff's cycle and the causal explanation suggested by Harvard professor Joseph A. Schumpeter, who tried to explain the existence of economic cycles by attributing growth to the fact that major technological innovations come in clusters. Kondratieff's postulation ended up being largely ignored for a variety of reasons. In the final analysis, however, the most significant reason for this rejection may have been the boldness of the conclusions drawn from such ambiguous and imprecise data as monetary and financial indicators.

These indicators, just like price tags, are a rather frivolous means of assigning lasting value. Inflation and currency fluctuations due to speculation or politico-economic circumstances can have a large unpredictable effect on monetary indicators. Extreme swings have been observed. For example, Van Gogh died poor, although each of his paintings is worth a fortune today. The amount and quality of artwork he produced has not changed since his death; counted in dollars, however, it has increased tremendously.

But in his book Ted proved the existence of 56-year cycles from quantitative observations of physical quantities using S-curves. Energy consumption, the use of machines, the discovery of stable elements, the succession of primary energy sources, and basic innovations were all reported in their appropriate units and not in relation to their prices. The cyclical variation obtained this way was more trustworthy than Kondratieff's economic cycle because they enjoyed the objectivity stemming from a scientific approach. In a way he scientifically sanitized the often-doubted Kondratieff cycle.

A graph with an S-curve in Ted's book was the spreading of AIDS. In the late 1980s and early 1990s the AIDS epidemic menaced society by its galloping rate of diffusion and the lack of effective medication. The only medical support available was a blood test for the HIV virus, and even that was not one hundred percent reliable.

Ted found data and made a graph of the evolution of the number of AIDS victims. He wanted to check whether the virus had a well-defined niche in society like so many other diseases. In his book diseases were treated as "species" that compete for victims. At any given time survival of the fittest dictates that the most "competitive" diseases claim the lion's share of victims. Toward the end of the twentieth century cardiovascular ailments claim the largest share, close to two-thirds of all deaths, with cancer second at about half that number. A hundred years earlier pneumonia and tuberculosis were running ahead of cancer. It seems that old diseases weaken and wither with time whereas young and more "competent" ones take over. Was AIDS to replace all previous diseases and to what extend?

The best S-curve Ted could fit on the AIDS data revealed a final ceiling of 2 percent of all deaths in the United States. A restricted place seemed to have been reserved for AIDS in American society; it was confined to the microniche of 2 percent of all deaths, in sharp contrast with alarmists' fears for the extinction of the human species.

But Ted was also dragged into the subject of AIDS from another angle. As a member of the editorial board of the journal *Technological Forecasting & Social Change* he was asked to review a difficult-to-read book entitled *Experts in Uncertainty* by Roger Cooke. He kept postponing the job and the book lay on his desk for over a month accumulating dust and making him accumulate guilt feelings. The task was particularly unpleasant because the book was what most people would classify as unreadable. It consisted of reviews of statistical methods evaluating the uncertainties assigned on probabilities of events, and was written in a professionally academic way. Despite the fact that Ted was attracted by the rigor and the thoroughness of the treatment, and despite the fact that the author assures the reader in the introduction that knowledge of advanced probability theory is not required, Ted kept putting off the job, dreading the numerous pages crowded with formulas and mathematical symbols.

His problem was resolved when a colleague came to him for advice. Ted's colleague had read an article in a popular magazine claiming that an AIDS test with a 5 percent false-positive identification rate may imply only a 2 percent chance that the person tested positive has indeed the HIV virus. How could that be? A test that correctly identifies a sick person 95 percent of the time is not a bad test. How can it be then that the chance of really having the virus is as low as 2 percent?

Ted's first reaction was disbelief. His second reaction was to tackle the problem in a physicist's way (that is, as if the problem had never been solved before) despite his colleague's pleas for looking up Bayesian logic formulations in a statistics book. It soon became clear to Ted that the frequency of AIDS casualties has much to do with the true probability for carrying the virus. If the epidemic is so well entrenched in a region that usually one in two people who go for an AIDS test carry the virus, a positive test becomes a reliable indicator. But suppose—more realistically— that only one in a thousand of those who become worried and go for a test actually carries the virus. Then, in a thousand individuals tested, only one will be really infected, while there will be fifty others who also test positive given that the test has a 5 percent false-positive identification rate. The probability then of actually carrying the HIV virus is only one in fifty, or 2 percent!

Ted played with these numbers further trying to get a better feeling for how the disease prevalence affects the reliability of the AIDS test; he was almost amusing himself with the observations when his colleague drew Ted's attention to the last paragraph of the article. A young American had tested positive on an HIV test, which they told him was 96 percent accurate. He assumed that his chance of having AIDS was 96 percent and committed suicide. The true chance however was only 10 percent.

Ted stopped his number crunching and picked up *Experts in Uncertainty*. He quickly found the appropriate section where the problem was extensively treated. In fact, he found several other practical applications of potentially popular interest. Suddenly, Ted was ready to write his review. Besides enumerating the topics treated by the author, and adding some positive comments about the book's scientific rigor and value as a reference manual, Ted felt the urge to make an appeal: "Could the author be persuaded to write a simpler version of selected useful topics from his book for the general reader?"

A few days later when Ted's daughter teased him by asking, "What would you do, Dad, if you found out that you tested positive in an HIV test?" Ted replied in a reflex, "I would rush to make another test."

In fact, by 1997 the World Health Organization had stipulated that all positive results must be confirmed by another test.

36 – Scientific Certification

The book project brought Ted back to CERN. His publisher, Simon & Schuster, wanted world-renowned scientists to affirm that the content of the manuscript was scientifically sound. Ted thought of Pascal, the Frenchman physicist leader of the group that he joined when he first went to CERN. Pascal had now become director of research. Ted had not seen him for several years, but Pascal was pleased to see Ted. When Ted told him about his book, Pascal quickly understood what was at stake. "They need scientific approval," he retorted. "Leave the manuscript with me."

A few days later he asked Ted to come for coffee and showed him the letter he had written. Ted was pleased and not only because Pascal assured Simon & Schuster that there were no scientific flaws in the manuscript. Ted's manuscript was breaking ground away from hard-core physics and he had been apprehensive about Pascal's reaction.

In fact, Pascal had a number of remarks and suggestions for improvements. The most significant suggestion was to introduce one more story to help the reader intuitively grasp the predictive power of S-curves ingrained in the symmetry of the S-shaped pattern, not a trivial task.

It was several days before Ted came up with such a story. Finally, he faxed one page of text to Pascal with a description of a rather distasteful fictional event, which however did display adequately—Ted thought—the crux of the mathematical equation that gives rise to the symmetry in the S-shaped growth pattern. The story went as follows:

> There is road construction going on next to a 13th-century church in central Paris. In front of a magnificent stained-glass window, red steel barrels filled with small stones line the work site. A teenage vandal passing by at night looks at the window, which contains over a hundred small pieces of glass. The bright streetlight reveals a black hole; one piece is missing. The contents of the barrel give him an idea, and he throws a stone at the hole. He misses, but his projectile knocks out another small pane; now there are two holes. He liked the sound it made, and so he throws two stones simultaneously. That brings the number of holes in the window to four. He decides to develop this game further, each time

throwing a number of stones equal to the number holes and giving up any pretense of aim. The random hits make the number of holes increase exponentially for a while (4, 8, 16, and so forth), but soon some stones fall through holes, causing no further damage. The vandal adjusts the number of projectiles per throw according to the rules of his game. But as the number of the glass pieces still intact diminishes, he realizes that his rate of "success" is proportional to this number. Despite throwing tens of projectiles at once, his hit rate is reduced by a factor of two each time.

Inside the church, under cover of darkness, a tourist is taking advantage of his sophisticated video-recording equipment in order to bring home pictures that tourists are not normally allowed to take. Among other things he films the complete demolition of the stained-glass window. Aware that he is trespassing, he does not interfere.

Back home he shows his friends the beautiful vitraux before and after. Faced with the dismay and indignation of the audience, he plays his tape backwards, and the masterpiece is reconstructed piecemeal. The audience watches the empty honeycomb-like lead framework become filled with sparkling colored glass, first one piece, then 2, 4, 8, 16, and so forth, until the exponential growth slows down. Someone points out that the growth pattern in the reconstruction is identical to the sequence of hole appearances in the original scene of vandalism.

But when Ted went to CERN to seek Pascal's feedback he learned that Pascal was not happy with the story. He said the topic of the story did not bother him but the mathematics was not accurately described. Ted tried to insist, but Pascal sent him away saying that he was very busy and instructed Ted to look into it more carefully.

Ted was frustrated. True, in the past, there had been occasions of disagreement between them on scientific issues in which Ted had been wrong more often than not. But this time he had really worked hard at it and he was sure of his answer. He went home checked and rechecked his story and finally faxed Pascal a short letter:

Dear Pascal,

Remembering several situations in the past where I insisted
on my positions that you later proved wrong, I have re-
doubled my efforts this time and can find nothing wrong with
my reasoning. Can we please make an appointment for you to
explain to me what I am not grasping?

Ten minutes later Ted's phone rang. It was Pascal, who admitted, "Yes,
you are right." He did not say anything else, but that was enough for Ted.
Physicists generally have difficulty admitting they are wrong.

Another one of the many requests Ted's publisher had was an endorsement
for the book's front cover by a "household science name; someone like
Asimov or Carl Sagan," they had suggested, or at least a Nobel Prize laureate
even if his or her name is not well known. Ted had immediately thought of
his thesis advisor, Jack Steinberger, who had in the meantime received the
Nobel Prize.

"Hello, Ted," Jack said when Ted visited him in Geneva where he lived.
"What brings you to our house?"

"Jack, I am writing a book," Ted said. "I brought you my manuscript for
your comments," and gave him the document entitled *Forecasting Destiny* at
that time.

In his late 60s Jack was far-sighted. He took the document by two hands
and stretched his arms as far as they would go. He began reading,
"Forecasting Dest…" Abruptly he gave it back to Ted turning his head the
other way. "You know me, Ted," he said, "I don't read this kind of stuff."

"But Jack," Ted protested, "You are my teacher, you signed my thesis,
and now I need your opinion on a book I am writing."

"I don't read this kind of stuff," he repeated himself raising his voice.

So much for the ace up Ted's sleeve as far as Nobel laureates were concerned. Jack was his best shot. He had so hoped to have an endorsement
from him. There was of course Rubbia that Ted had met while at CERN but
he had now become director general and was impossible to see. So Ted
decided to try Simon van der Meer, the Nobel laureate at CERN, who shared
the Nobel Prize with Rubbia. He was an applied physicist who contributed
crucially in the construction of the particle accelerator that Rubbia used for
his experiment.

37 – A Down-to-Earth Nobel Laureate

Van der Meer had recently retired but had remained in the Geneva area, so Ted called him at home. Ted reminded him that they had met years ago when they were both at CERN and told him that he wrote a book on which he wanted his opinion. Van der Meer asked for a copy of the manuscript so Ted brought one to his home. A mere two days later van der Meer called Ted. "I have read your manuscript," he said, "why don't you come here to discuss it with me?"

When Ted later realized how detailed and meticulous a reading van der Meer had done, it was clear that he must have done nothing else during those two days.

Van der Meer was an applied-physics Dutchman with practical thinking and nothing flamboyant; how different from Rubbia, Ted thought. Van der Meer skipped all unnecessary preliminaries (how are you, would you like a drink, etc.), invited Ted to sit next to him, and opened up the huge binder with Ted's manuscript.

He began by saying that he found the book very interesting and thought provoking. But then he proceeded to explain that he had carefully examined the document and had many comments to make. Indeed there were pencil marks on almost every page. As they turned the pages Ted became worried about the kind of problems his objections might create, but soon realized that there was no seriously bad news.

In one graph he argued that the S-curve Ted had drawn was not quite symmetric. In another he wanted to see error bars on each data point. On another page he argued that learning how to play tic-tac-toe does not really follow an S-curve, and despite his English not being his mother tongue, he managed to find spelling errors and typos.

But his most vehement objection concerned Ted's claim that the prehistoric people who built the Stonehenge monument in southern England had significant knowledge about periodic astral phenomena that we may not yet fully grasp today. "It is absurd to imagine that primitive people knew more than we do," he concluded.

When they arrived at the end, he took a deep breath and asked, "Now what do you want from me?"

"Well," Ted began "with your permission I would like to authorize my publisher to put an endorsement from you on the book's cover jacket."

"Oh, no!" he reacted. "I don't like that. You cannot trust what journalists and publishers may come up with. They are likely to say that a Nobel Prize winner claimed this, that, or the other. I really don't like that."

Ted gathered his courage, "But Mr. van der Meer, you said that you found the book enjoyable, interesting, well-written, and thought provoking, didn't you?"

"Yes I did," he replied "but I also found it ... hmm... what's the word..."

Ted knew he was looking for some unflattering word so he did not volunteer anything. A few seconds later, van der Meer had found it.

"Controversial," he uttered. "This book is controversial," he repeated.

Ted was relieved. "Yes, of course, we will add the word controversial," Ted assured him. "There will be only these five words, not one more or one less. I'll request the proofs to be sent to you before publication for you to check the exactness of the typesetting," Ted persisted.

Van der Meer was momentarily pacified but soon came back with more objections. "There is something else. I cannot possibly have my name on a document that supports astrology. There is a section where you argue in favor of astrology and I will not go for that."

He was obviously referring to the section in the book entitled *How to Defend Astrology without Risking Your Reputation* in which Ted argued that the relative positions of the moon, the sun, and the earth could be influencing the climate on earth, which in turn could be influencing human affairs. So Ted tried to argue.

"But in the section you are referring to I am not defending astrology as such," Ted began defensively when van der Meer interrupted him. "Yes, yes, I know what you are saying and there is nothing wrong with that, but *it appears* that you are endorsing astrology. I simply do not want my name on a document that uses the word astrology at all."

Ted sensed the man's aversion to this word and without hesitation he offered, "I will have the word removed. I hereby assure you that there will be no mention of astrology in my book."

Here is a central excerpt from the section van der Meer objected to. It enjoys *bona fide* scientific rigor.

• • •

The people of Stonehenge, long before Christ, through observation of celestial phenomena, had become conscious of a 56-year period.

Astronomers have long been aware of celestial configurations that recur in cycles. The cycle of Saros, known since antiquity, is based on the fact that identical solar and lunar eclipses occur every eighteen years, eleven days, and eight hours but will not be visible at the same place on the earth. The cycle of Meton, which has been used in the calculation of the date of Easter, is based on the fact that every nineteen years the same lunar phases will occur at approximately the same time of the year.

In fact, lunar eclipses recur and are visible at the same place on the earth every 18.61 years. Therefore, the smallest integral-year time unit that allows accurate prediction of eclipses at the same place is nineteen plus nineteen plus eighteen, a total of fifty-six years.

Lunar and solar eclipses figure prominently in superstition, but their importance goes beyond that. Biological effects, including strange animal behavior, have been observed during eclipses. The mere effect of having the three celestial bodies on a straight line provokes exceptional tides. There are periodicities of fifty-six years on the prediction of tides. But the fifty-six-year period concerns not only eclipses and the alignment of the earth, moon, and sun on a straight line. *Any* configuration of these three bodies will be repeated identically every fifty-six-years. Possible effects on the earth linked to a particular geometrical configuration will vary with the fifty-six-year period.

There is one more astral phenomenon, completely independent, which displays a similar periodicity: sunspot activity. For centuries astronomers have been studying spots on the surface of the sun. Cooler than the rest of the sun's surface, these spots can last from a few days to many months, releasing into space huge bursts of energy and streams of charged particles. The effects of sunspot activity are varied and continue to be the object of scientific study. What is known is that they are of electromagnetic nature and that they perturb the earth's magnetic field, the ionosphere, and the amount of cosmic radiation falling on earth from outer space.

We also know that there is a regular eleven-year variation in sunspot intensity. Sunspot activity reaches a maximum around 2023, 2012, 2001, 1991, 1980, 1969, etc. During a maximum, the overall solar output increases by a few tenths of 1 percent. The corresponding temperature change on the earth may be too small to be felt, but meteorologists in the National Climate Analysis Center have incorporated the solar cycle into their computer algorithms for the monthly and ninety-day seasonal forecasts. Every fifth period (five times eleven equals fifty-five years) the timing of the sunspot variation will be close enough to resonate with the fifty-six-year cycle. Moreover, in the three-hundred-year-long history of documented sunspot activity, we can detect relative peaks in the number of sunspots every fifth period.

All these celestial influences are probably too weak to significantly affect humans directly. They may affect the climate or the environment, however. There are regularly spaced steps on the continental shelf, and darker circles have been observed on centenary tree cross-sections, with a comparable periodicity. Both indicate environmental changes, and if the environment is modulated by such a pulsation, it is not unreasonable to suppose that human activities follow suit. In fact, observations have linked climatic changes and human affairs.

● ● ●

Van der Meer was not entirely wrong being excessively cautious concerning kinship to astrology. After the book's publication with all references to astrology removed as promised, Ted nevertheless received a phone call one day from a man who spoke English with an Indian accent and presented himself as a researcher and author of esoteric literature. He congratulated Ted for a wonderful work and proceeded to explain to him how happy he was that Ted had given scientific backing to astrological divinations. When Ted argued that it is not what he had done, the man refused to acknowledge what Ted was telling him; he thanked Ted again profusely and bid him goodbye.

Van der Meer's vehement objection to astrology was something Ted could sympathize with. He had done worse himself in his freshman year at college. When he first registered at Columbia he received quite a few credits of advanced standing for work he had done in high school, particularly in chemistry and the humanities. Consequently his course advisor suggested elective courses to fill the gap, and being in physics, he proposed a course in astronomy. But astronomy and astrology shared nonscientific connotations in young Ted's mind and he refused. He took chemistry instead despite the fact that he would be repeating material he had already studied. He never forgave himself for his narrow-mindedness because he missed what turned out to be his only chance to become seriously acquainted with astronomy.

38 – Fallout

Following the book's publication there was considerable fallout. One situation brought Ted face to face again with Carlo Rubbia. It was a lunch invitation by the American Ambassador.

Morris B. Abram, the American ambassador to the United Nations who heard about Ted's book on a local radio program contacted Ted asking to buy a copy of the book. When Ted sent him a complimentary copy, he replied that he would like to have Ted for dinner at some future occasion.

The ambassador's official invitation came a few weeks later. Notable personalities of the region and Ted were invited for lunch at the American Embassy. The list included Nobel physicist Carlo Rubbia, who was the guest of honor.

The American embassy in Geneva was situated at the top of a hill and it was fortified like an army outpost even in those peaceful early days of the Clinton administration. But being the ambassador's invitee permitted Ted to drive and park inside the compound unhindered. The dinning room, on the building's sixth floor had large windows providing a spectacular view of the city of Geneva and its lake.

Following cocktails they were seated at a large round table and the ambassador went around the table making introductions. There were a couple of professors in economics, the dean of the Graduate Institute of International Studies, and of course, Carlo Rubbia, whose dominant personality proved disastrous for the ambassador's plans for this gathering.

Early on, the ambassador said that he wanted to suggest a theme up for discussion during the meal: Nature versus Nurture. Apparently he was still under the influence of an article on that topic in the current issue of *Time* magazine. But Rubbia monopolized the conversation from the beginning with stories of experimental physics that he considered amusing. From time to time the ambassador would reiterate, "And if we were to come back to our topic…" Rubbia seemed impervious to the ambassador's pleas.

Finally the ambassador made a gallant effort to force a change of the subject. Rubbia paused for a few seconds and then continued, "Well, there was this man called Klein, who had no education but an amazing ability to carry out complicated mathematical operations in his head."

Ted remembered the man. He was able to carry out in his head not only tedious multiplications and divisions involving figures with many digits but also calculating square roots, third roots, integrals, and many other such brain-twisters without ever making mistakes. A real phenomenon! Back in the 1950s computers were not easily accessible so Klein was given a job at CERN as a … calculator! He would walk up and down the halls of the laboratory and answer questions of physicists on arithmetic operations.

"Well, what do you think of that for a case of nature versus nurture?" concluded Rubbia. "This guy had an incredible talent and what came of it? Stupid mechanical calculations. Moreover, it turns out he was a homosexual!"

Ted felt compelled to interject that Klein had succeeded nevertheless to secure a permanent employment contract at CERN, a position coveted by many highly educated scientists.

By this time Ted had abandoned all attempts to bring up for discussion ideas from his book. During cocktails he had mentioned to Rubbia that his life curve was in Ted's book together with the life curves of other personalities from the arts and the sciences, but this did not distract Rubbia from the subject with which he was preoccupied at that moment.

Ted realized once again how difficult it is to have an effect on someone whose "momentum" is large.

On another situation Ted proved more effective. This time the fallout was from the book's Greek edition. The day had been full of excitement and emotions. It began in the morning with a press conference where the book was presented to the public and to journalists. Then there was a long lunch for a dozen friends and relatives in a distinguished Athenian restaurant. Late in the afternoon Ted had to give a talk at the University of Athens from where he was whisked to the television studios to appear on national TV. A woman journalist who presented the prime-time news every night interviewed him for half-an-hour. Her questioning was rapid but friendly. At the culmination of a day full of intensive and rewarding activity Ted was in a euphoric state of top performance. His answers were as rapid as the questions and more to the point. Among other things she asked him.

"We have seen much 'rose' literature recently on scientific topics. Would you say you are doing some kind of popularization of science with your book?"

Ted did not know what she meant by "rose" literature. He assumed she referred to vulgarization of science. "Not at all," he challenged her. "I like to think I am doing some kind of commercialization of science."

As they had been talking in Greek she continued, "Your Greek is excellent but you have lived so long abroad you may be using the words inappropriately. Are you aware that 'commercialization' carries bad connotations, something like 'degrading' and 'cheapening'?"

"I disagree," Ted insisted, "there is nothing degrading in extracting usefulness and even profit from science. After all, that is its *raison d'être*. Science is not meant to be locked up in ivory towers and be accessible only to a select group of people. It is meant to deliver value by all possible means, including ways not anticipated by scientists."

His answers came spontaneously. He managed to conclude with a punch line forecasting an upcoming glorious period for Greece. He could practically "hear" the people applauding in their homes in front of their TV sets.

It was not Ted's only appearance on TV, but it was by far his best. The intensity of emotions and excitement during that overly charged day had resulted in an exceptional performance on his part.

39 – Growth Dynamics

Back in Geneva DEC had embarked on a declining S-curve. The Management Science group now seemed to be a luxury that DEC could no longer afford. As employees scrambled for the reduced number of job positions, Ted witnessed first-hand the beneficial aspect of hard times. People became innovative and entrepreneurial. New ideas surfaced every day. He himself thought more about the life-cycle model, mapped it on the cyclical variation of the four seasons, and deduced from it optimal behaviors and strategies that depend on the season one happens to be in. He was preparing for a consulting practice of his own around the slogan: where are you on the curve and what to do about it. When the moment came, he launched his own consulting company, Growth Dynamics. It specialized in new ways of thinking in business strategy, forecasting, the stock market, and decision-making in general.

Ted was scared but also excited to be working on his own. He approached it as another case of "rejoice rejoice we have no choice." But he did not hesitate to knock on the door of a big consulting firm when he saw an intriguing advertisement in *The Economist*. Under the title "Focus on the Horizon" the consulting firm Ernst & Young was looking for "thought leaders with demonstrated skills in long-range planning, applied research, and strategic consulting." They asked for a brief letter outlining experiences and a *Third Horizon* topic (meaning concerns that may come up in five years' time) to be explored.

In his fax to Ernst & Young Ted included the following:

> I have recently established my own firm, Growth Dynamics, in pursuit of some compelling ideas that I outline below.

> The preachings of management consulting gurus have been ever-changing and are often contradictory. They come and go like fads. We have swung from leadership to empowerment, from excellence to innovation, from total quality management to business process re-engineering. It is disconcerting to think that there is an endless stream of breakthroughs in the theory of doing business. My aspiration has become a business theory that can last the test of time all encompassing the life-giving

variation in strategic thinking evidenced by the gurus. A cross-discipline viewpoint combined with an understanding of how growth processes evolve over time produces a fascinating Big Picture. The contributing elements range from natural laws to the doctrines of Darwinism (and postdarwinism), chaos, classical sciences, and quantum mechanics. The future horizon extends to the decade of the 2020s, in all probability the next global economic boom and climax of prosperity.

I want to go beyond where Charles Handy stops in *The Age of Paradox*. In his book, he advocates uninterrupted vigilance for the turning point. The beginning of the downturn, he warns, is to be assumed imminent at all times, because "there is no science for this sort of thing."

But the idea of stubbornly and continuously preparing for disaster insults my pride as an intelligent human species. It is as inefficient to continuously gear up for disaster as it is dangerous to assume that the good days will last forever. I do have in my disposal a science that will anticipate the next turning point with *sufficient* accuracy for *just-in-time* action.

My approach is both theoretical and practical, and covers two main topics:

A. **The Power of Strategic Evolution.** Business is like nature in that it goes through seasons, and so do the successful business strategies. One's first priority should be to determine objectively and in a timely manner the present business season and duration. The next thing is to act in accordance to the spirit of that season.

B. **Competition Management.** Despite the success of biological models and lessons from Mother Nature, survival in the marketplace differs from that in the jungle by the fact that in the former the roles of predator and prey are not genetically predetermined for the competitors. In the marketplace the nature of competition can be changed via the right decisions at the right time.

> I am pursuing the development of interactive computer-based software systems, one on Competition Management that offers quantitative guidance on pricing moves, on the advertising message and investment, and on actions destined to influence the competitive dynamics. A separate tool handles the determination of the business season, its duration, the timing of the turning point, and the measurement, monitoring, and assessment of one's strategic performance. A welcome byproduct is the evaluation of one's intuitive versus one's rational appreciation of the business situation.

Ernst & Young took the bait. They called Ted and arranged for a long interview by telephone. During the interview they asked for documentation, copies of Ted's recent writings. He could only fax to them several articles that he had published, most of them in *Technological Forecasting & Social Change*. That was academic rather than business material and did not do the job. He did not hear from them again.

However, the incident motivated him to arrange all the material from his articles in the form of a book. He set aside all other work and within six weeks he had produced a manuscript that was soon published as a business book.

The new book made an effort to reconcile the old trend (equilibrium economics) with the new trend (complexity economics) in such a way as to transcend the passage of time, just as hard science introduced modern physics without negating classical physics. Knowledge is cumulative; it cannot be forgotten. The craze about chaos studies in the 1980s raised much hope but did not succeed in cracking the stock market. Then the focus passed to such notions as complexity, self-organization, symbiosis, and co-evolution. But again there were no miracles. It must be recognized that there can be no single panacea-type theory like a miracle drug. In a chapter entitled "The Evolution of the Evolution" Ted integrated S-curves with chaos. Sustained growth does not follow a straight line, or an exponential, or chaotic behavior, or some other single pattern. Sustained growth goes through well-defined natural-growth steps. The emerging picture alternates between states of order (rapid growth) and states of chaos (slow growth) as old niches fill up and new ones open up.

40 – "If winter is here, can spring be far behind?" *Percy B. Shelley*

The seasons metaphor that Ted employed to derive strategies for the four stages of the growth cycle had much success with his early customers. His slogan became: Is your business appropriately "dressed" for the season it is traversing? He used this approach for anything that grows in competition, be it product, market, professional career, or personal relationship.

S-curves are known to cascade with a new one beginning where the last one leaves off. New products replace old products just as new technologies replace old technologies. A well-known marketers' utopian pursuit is the strategy of product replacement timed so as to avoid any slowdown in the growth of their overall sales revenue. Evidently, launching products too closely together (the case may be argued for new Microsoft Windows operating systems) may frustrate customers and/or lead to "cannibalization" of their own market when the new product robs sales from the old one. On the other hand, delaying the launching of a replacement product may create a vacuum in a vendor's offerings and result in loss of customers to the competition. So the question becomes when is the optimum time to launch a replacement?

But also in careers and personal relationships one may find successions and replacements as a new one begins when the old one ends. The question can be generalized to: when is the right time to introduce change in an ongoing natural-growth process? No one wants to tamper with something that works well, but how old should become a product before its replacement is launched?

The criterion for optimum timing in replacements can be found in harmonic motion and not only because the concept of harmony implies goodness. Regularly spaced product life cycles produce a landscape pattern suggestive of a sine wave, and large-scale cyclical phenomena, such as the Kondratieff economic cycle. The sinusoidal wavy pattern is characteristic of the pendulum's harmonic motion.

Ted studied the conditions for what he called a "harmonious" succession of S-curves, where the sum of their life cycles perfectly matches a sinusoidal wave. It turned out that optimum introduction of change occurs when 3 percent completion of the new process coincides with about 90 percent completion of the old process. This then can serve as a quantitative rule for the perennial quest of *just-in time* replacement.

But harmony in replacements does not maintain the high overall growth rates that marketers wish for. The overall envelope undergoes peaks and valleys and the business seasons feature summers and winters. The dip in revenue during the replacement phase is natural and inevitable. More than that, it is desirable. This dip plays a significant role in triggering new growth, just as pruning the roses ensures healthier blossoms for the next season.

TWO NATURAL-GROWTH PROCESSES CASCADING "HARMONIOUSLY"

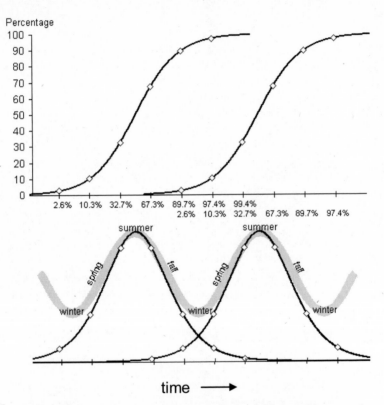

The two S-curves represent two consecutive natural-growth processes. The corresponding life cycles below show how their rates of growth go through "seasons." The sum of life cycles cascading in this way matches perfectly the thick gray line, which is a sinusoidal (harmonic) wave. The little rhombs delimit the seasons.

In personal relationships the winter dip corresponds to a period of mourning. Much has been said by psychologists and couple therapists on the importance of acknowledging the separation, going through a mourning process before moving from one relationship to another.

In his second book Ted tabulated a large set of behaviors, each one best suited for a particular season. Becoming conservative—seeking no change—is appropriate in the summer when things work well; this is the time to strive for excellence and total quality management. But excellence drops in second place during the difficult times of the winter when fundamental change must take place; one should now become entrepreneurial and innovative.

Learning and investing are appropriate in spring, but teaching, tightening the belt, and sowing the seeds for the next season's crop belong in the fall. In spring the focus is on *what* to do, whereas in fall the emphasis shifts to the *how*. The former appears early in the growth process, the latter late. Take the evolution of classical music, for example. It can be visualized as a large-timeframe S-curve beginning sometime in the fifteenth century and reaching a ceiling in the twentieth century. In Bach's time composers were concerned with *what* to say. The value of their music is on its architecture and as a consequence it can be well interpreted by any instrument, even by simple whistling. But two hundred years later composers such as Debussy wrote music that depends crucially on the interpretation, the *how*. Classical music was still "young" in Bach's time but was getting "old" by Debussy's time. Mozart, Beethoven, and Chopin enjoyed, but also shaped, classical music's "summer" season.

When you hear people say that they need to focus on the how, you can understand that they are talking about something that is getting old.

THE EVOLUTION OF CLASSICAL MUSIC

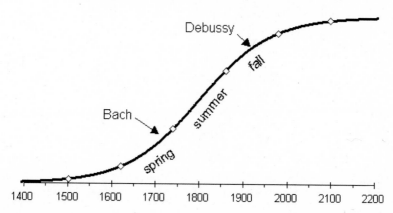

Winters are full of change and change can lead to death. A species population that undergoes large fluctuations may hit zero at some point, in which case the species will become extinct. "If the pain keeps changing/moving, it will eventually go away," had been the advice of a wise rheumatologist that Ted consulted once about a persisting pain on his shoulder.

But winters are followed by springs. "Death normally comes in winter but it doesn't have to be this winter," became Ted's slogan. He offered guidelines on how to maximize one's chances for what he called "A Second Lease on Life." That is what Hitchcock achieved by branching into television movies when his feature-movie career slowed. That is what Mozart missed when he composed his dissonant quartet in 1785. Subconsciously aware of the fact that his creativity had entered a season of fall, Mozart tried to branch out into a new type of music, which was, however, ahead of its time. Contemporary music lovers could not adjust to the kind of music that became acceptable more than 150 years later, and so Mozart died during that winter of his productivity curve.

Besides providing him with material for amusing stories, the seasons model enabled Ted to properly handle the difficult-to-accept predetermination ingrained in S-curve forecasts. Once a growth process is well under way, there is no way to avoid its completion under *natural* conditions (natural conditions are the type of conditions encountered during the historical window of the data). This message did not go down well with marketers, who thought they had the right to exercise their free will all times.

Ted's advice was that new directions should be set during "winters," when change is rampart. But to give a new growth process a realistic chance to survive, the changes introduced should be tantamount to creating a new "species." Minor improvements on a species already on a phasing-out trajectory are not sufficient to provoke a turnaround. For Mozart to compose another clarinet concerto would have been business as usual, but to compose dissonant music was indeed tantamount to creating a new "species," if an ill-timed one.

In contrast, during "summers," the established course should not be tampered with; only fine-tuning should be attempted. Leaders do not have much choice during this business season. They are forced to follow the established course. This season is *par excellence* the time for total quality management and the pursuit of excellence. Of course, decision makers maintain their free will to make a mistake at all times!

Ted's approach and the associated software he developed could be used not only on business situations but also on every activity that undergoes a cyclical variation such as that of the bell-shaped life-cycle pattern. A section in his new book was entitled "What to do when you do not know what you are doing." It included a version of the philosopher's stone.

In the alchemists' tradition, the philosophers' stone was a desired object that would tell you the right thing to do, whatever the situation.* The version Ted offered was a procedure that called for taking three steps before responding to change, be it in your business, or private life:

1. To the best of your ability make a guess about what is the season you are in, where in the season you are, and how long it will last. (Ted had built sophisticated tools to help his customers do that.)
2. Go back two seasons and identify activities that proved successful then.
3. Opt for the opposite of those activities. Make sure your actions are in harmony with your season's attributes as assigned by the seasons metaphor.

---

* It is a hasty simplification to say that alchemists' main goal was to find a way to turn base metals into gold. The alchemical work was symbolic and esoteric, not unlike Gurdjieff's. It aimed at ultimate knowledge and eventual "awakening." The philosophers' stone was a key substance in achieving that goal.

41 – Scientifically Sanitized Biorhythms

At some point, and thanks to the Japanese edition of his book, Ted found himself in Tokyo where he visited the famous Akihabara shopping district, which specialized in electronic goods. There in the midst of a sea of gadgets he saw a device that looked like a small calculator but displayed biorhythms when one typed in his or her date of birth. Ted's attention was caught by the three sinusoidal patterns drawn on the small screen. They looked like sequences of life cycles, and a little arrow pointed out where present is. Below the drawing appeared text referring to the particular configuration at present and gave an analysis and advice that seemed general, vague, and arbitrary, like that of mediums and astrologers. Ted immediately thought that his approach of seasonal behaviors could yield more valuable and more defensible comments about what one should do given where one is on a life-cycle pattern. The problem was that Ted never took biorhythms seriously. He had them classified into his weirdoes cabinet in a drawer with many question marks on it.

Biorhythms are three biological cycles that presumably govern human behavior: the physical, the emotional, and the intellectual (or mental) cycle. All three are set in motion simultaneously at birth and oscillate regularly with periods of 23, 28, and 33 days respectively. Even though they all begin at zero, the different periods cause the three cycles to go out of phase rather quickly, and produce an infinite number of configurations over time as one's physical, emotional, and intellectual states go through good and bad periods. Each configuration then becomes subject to interpretation, forecasting, and advice.

Ted had no problem with the idea that one's performance goes up and down. He was even willing to accept the hypothesis that the three waves may indeed be characterized by periods of 23, 28, and 33 days. But he could not accept the assertion that biological clocks keep ticking precisely for years and decades after birth. A wave with a period of 28 days goes through more than 1,000 cycles during a lifetime. The slightest deviation from the numbers 23, 28, and 33 would quickly result in very different configurations for the three cycles.

But there was an easy way to solve that. One of Ted's software packages featured six different science-based methods for determining where one is on a life-cycle curve, and that did not depend on some periodic variation being stable over decades. The birth date was in fact no longer needed and the seasons metaphor could provide plenty of analysis and advice.

The idea of sanitizing biorhythms scientifically excited Ted. When he got back to Geneva he modified the simplest one of his methods for positioning oneself on a life cycle and adapted it to the biorhythms problem. The software was simple; it made use of the mathematical concept of the second derivative, which people feel in their guts like a car's acceleration. There were three intuitive questions to answer by clicking on the appropriate box. They referred to one's disposition emotionally, intellectually, and physically. The first question was, "Do you feel optimistic or pessimistic about your prospects during the next few days?" The second question was, "Is the optimism (or pessimism) great or small?" The third question was whether the optimism (or pessimism) was increasing, stable, or decreasing with respect to a few days ago. That was all that was required.

On the first opportunity he presented it to his daughter, now a young lady susceptible to predictions and advice about her future. She answered as follows:

Do you feel OPTIMISTIC or PESSIMISTIC about the evolution of your physical, emotional, intellectual state during the next few days?

	OPTIMISM			PESSIMISM		
	Improving*	Stable*	Deteriorating*	Improving*	Stable*	Deteriorating*
LITTLE Refers to OPTIMISM, PESSIMISM						Emotional
MUCH Refers to OPTIMISM, PESSIMISM		Physical		Intellectual		

* Compare to how you felt a few days ago.

When she clicked on the three squares, the following diagram appeared on the computer's screen. The gray band indicated her present positioning and its width the uncertainty. The little crosses delimited the stage of growth (referred to as seasons in Ted's consulting practice) and the numbers on the horizontal axis were days.

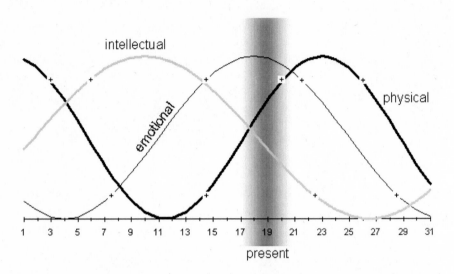

Now came the moment for the interpretation and advice.

"We see that at present you are physically in late spring, emotionally in summer, and intellectually in a fall season," began Ted. "During this week your emotional aptness and performance is at a peak, intellectually you are tired, and you should invest a little in your physical condition. In other words, you can trust, rely upon, and draw from your emotions, should spend some money on your appearance and generally take care of your body, and call upon your rational self only for the basics.

"In fact, you could benefit from re-examining the way you make rational decisions. This can be done via a comparison either with older times when your performance was good, or other individuals you know who seem to be doing well in that respect. It seems that you know what to do, but have hard time putting in into practice. (This is the do-as-I-say-not-as-I-do syndrome.) Concerning academic activities, it is a good time to teach but not a good time to try to take exams.

"Concerning business decisions, you should take everything into account but make the final decision instinctively. Try to win people's sympathy rather than arguing with them. If you need to exert pressure, you have a better chance to succeed either by scaring people or by getting them excited, but not by persuading them.

"On the personal relationship level, it is advisable to indulge in feelings this week, and postpone physical performance to next week if possible. In the meantime, you should rest and eat well. You need to avoid thinking too much, do physical exercises (consult specialists like nutritionists, kinesiotherapists and the like), and let your emotions free (buy your boyfriend a painting rather than a book, take him to a concert, rather than to a conference). Explore and try to integrate more than one emotional medium: poetry, music, drama, jealousy (but take no risks), presents, children, animals, pets, etc."

Ted liked the way his advice sounded and could have continued like this for a long time. His interpretations were plausible, coherent, and could be further elaborated. Moreover, there was nothing special about the particular configuration of his daughter's biorhythms. *Any* configuration of the three cycles would be amenable to interpretations of this kind.

In fact even if a person does not believe in biorhythms at all, he or she could benefit from this exercise. There is a minimum intrinsic usefulness in such an activity, just as there is in the Chinese I Ching, the tarot, and the inkblot psychological tests. They all serve as a means for projecting wise subconscious judgments onto a given situation.

His daughter had an idea. "Why don't you do this professionally," she said. "People pay real money to fortunetellers. There is a big market out there, and if you sell it as a scientific method, go on TV and say that you are a physicist and were at Brookhaven and CERN, you will beat all the competition from run-of-the-mill mediums. Just think, you could win over clients like presidents Ronald Reagan and Francois Mitterrand." She was obviously referring to the well-known penchant of some world leaders for astrological divinations.

Her suggestion amused Ted but he could not consider it seriously. Not because it was unfounded, but because he couldn't break away from science by that much. Part of him would have to die. Having partaken in *real* knowledge is an irreversible process—as Michel de Saltzman had put it concerning involvement with Gurdjieff's work. One may wander off the beaten track, but one can no longer make *tabula rasa* by adopting behavior entirely incompatible with what one lived by for the most part of his or her life. People are known to even opt for death at moments of profound deception, like the captain whose boat sinks.

Nevertheless, Ted buried his modified software program in an obscure corner of his Growth-Dynamics website. Under the heading "Extra Curricular" he introduced a link to a "Bio check-up" where visitors could answer a questionnaire, position themselves on a biorhythm chart, and get insights in accordance with the seasons metaphor. He went as far as to incorporate in his program a calculation of an "effective birthdate", which would yield the same configuration of the three waves. Visitors could then use this date—instead of their own date of birth—in the numerous traditional biorhythm generators, in order to obtain readings that corresponded more accurately to their present state.

42 – A Leadership School Experiment

The fallout from Ted's books continued for many years and even though he did not become rich he was presented with many interesting situations. One of the most spectacular ones began with an e-mail he received:

> "I am the dean of a leadership school in Monterrey, Mexico, and I would like to invite you to give a course on forecasting. The intent of the course is to provide a broad understanding of effective ways to look into the future—be it on social, financial, or technological issues—emphasizing more the thinking and less the techniques, the way you approach forecasting in your books."

The invitation thrilled Ted. It would be refreshing to teach for once with a *carte blanche* on the curriculum and to do away with the tedious traditional business-school methodologies.

The Mexico experience turned out to be more exciting than expected. Carlo Brumat, another Italian physicist-turned-business professor, had been given the task to create a leadership school for Mexico's promising young men and women. Alfonso Romo, a wealthy Mexican industrialist, had put up the funds for this school experiment. Brumat sought out renowned scientists and academics around the world who in addition to their expertise on a particular topic would also possess a bigger-picture viewpoint. They would be flown in, live in, teach, and be available to the students throughout their course. The courses would be short and intense. The entire calculus was a three-week course. Ted's would be one-week course. The students, about twenty-five of them, had been handpicked. The faculty had only scant interactions between themselves during the academic year, but they would all be brought together again for the better part of a week during the summer for graduation ceremonies and faculty meetings. It was during these summer sessions that Ted met the most intriguing world specialists.

After his first year, Ted knew what to expect. Each time he would meet new faculty members there would be two questions in his mind: What was the subject matter they taught and what was their special added value? He remembers hesitating after each introduction, making small talk while fishing for the juicy part.

"Nice to meet you, Keith. And what do you teach?"

"I teach calculus."

"Where are you from?"

"Well, I am a professor of mathematics at Stanford," and after a pause, "but I am also appointed to the chair of social mathematics that is in the process of being funded."

From then on, the discussion would revert to "what is social mathematics," and so on.

Ted enjoyed teaching his course. Moreover, receiving $10,000 for talking to a captive audience of smart young men and women for a week about his favorite subjects with no strings attached was enough to spoil him. To top it off, during the low winter seasons Air France made frequent promotions for the Concorde and Ted was able to upgrade the New York leg of his business-class ticket simply by asking.

He considered himself fortunate to have set foot on this airplane before it was discontinued. The French-British Concorde became the object of one of the case studies he taught. It demonstrated the phenomenon of precursors.

The evolution of natural-growth processes may display deviations around the smooth S-shaped pattern. When the analytical mathematical expression of the S-curve is cast into a discrete formulation, instabilities emerge at both ends of the growth process. The onset of these instabilities includes the precursor, the catching-up effect, and the ceiling overshoot.

In real-life situations we cannot see the early oscillation completely, because negative values have no physical meaning. We see, however, a precursor followed by a quiet period, then an accelerated growth rate, and finally an overshoot of the ceiling. These features correspond to real phenomena. Accelerated growth is a catching-up effect, usually attributed to pent-up demand. The overshoot is a typical introduction into the final steady state. As for the precursor, it is often considered a fiasco, unfairly so.

An in-scale drawing of first-order deviations from a natural evolution of supersonic travel reveals important long-term forecasts.

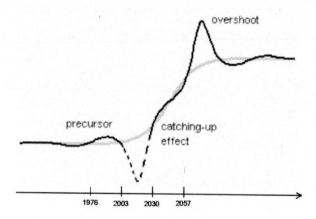

The Concorde enjoyed much publicity and popularity. It made a cultural dent and demonstrated the public's appreciation and need for high-speed travel. Its lifetime was limited for technical reasons. But it constitutes a precursor in the supersonic-travel growth process. Following the Concorde there may be another 27 years or so with no commercial supersonic planes. However, once a new-fuel technology becomes available (probably based on hydrogen or natural gas in liquid form), growth in supersonic travel will be rapid because of pent-up demand. It may lead into a supersonic "craze" (overshoot) around 2060 and finally stabilize at a lower level toward the end of the twenty-first century.

But extravagant ventures such as the DUXX Graduate School of Business Leadership cannot last very long. The tuition paid by the students was nowhere near sufficient to cover the expenses for running the school. Following the stock-market bubble burst in 2000, the affairs of Romo progressively deteriorated, forcing the closure of the school in 2003. In total, less than two hundred MBLs (Masters in Business Leadership) were awarded. Ted was among the lucky few who participated in this unique experiment.

43 – Insights into Fluctuations

When S-curves cascade they are separated by low-growth periods (winter seasons) during which a state of chaos appears, reflecting turbulent times. There is both theoretical and practical evidence for the appearance of chaotic fluctuations at the end and at the beginning of an S-curve. But during one of Ted's talks someone in the audience objected to Ted's schematic representation of sustained growth as a sequence of cascading S-curves with interspersed chaotic intervals.

"There may also be chaotic fluctuations all along the rising part of the S-curves," he argued. "They are simply less visible because they are masked by the pronounced upward trend."

To drive his argument home the man walked up to the blackboard and drew an S-curve with a trembling hand imitating an old man whose hands are shaking. Indeed on the curve he drew there were fluctuations everywhere but they seemed more pronounced at the extremities of the S-curve.

The drawing at the blackboard gave Ted time to prepare his answer.

"Granted, there may fluctuations all along the S-shaped pattern," he admitted. "But there is a significant difference between fluctuations during the steeply rising part, and fluctuations at the top and at the bottom. During the fast-rising trend fluctuations can be considered 'benign' as they generally correspond to reality, whereas at the top and at the bottom they are 'true-blue' chaotic as they will never become realized. Consider the fluctuations I will highlight with a little circle. I will indicate the corresponding level on the S-shaped pattern with a large dot," he said and proceeded to add little circles and large dots on the drawing on the blackboard.

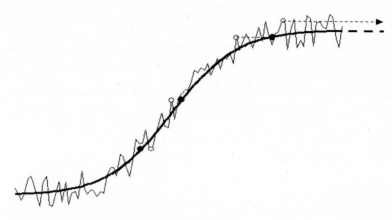

"During the steeply rising part, the distance between circle and dot is small, but it becomes bigger as we approach the ceiling. At the ceiling this distance becomes infinite. A fluctuation during the rising trend corresponds to a phenomenon that was naturally expected either a short time ago or in the near future. The only 'abnormality' introduced by such a fluctuation is that of a precocious or a belated appearance. But the same-size fluctuation at the ceiling reaches a level that will never be achieved by the natural-growth process, and thus has zero chance of being *naturally* realized. Such a fluctuation can be considered as unnatural.

It would be simply *unusual* if it snowed in Greece in October, but it would seem *unnatural* if it snowed in the Sahara. During irregularities events occur earlier or later than their natural time, but during chaotic events they have no corresponding time in which they could seem natural."

Without realizing it, Ted had come up with a procedure for distinguishing between natural fluctuations (i.e., simply statistical fluctuations) or unnatural ones (i.e., chaotic). Naturalness is intuitively associated with the probability of being realized by the natural-growth process depicted by the smooth S-shaped pattern. A downward fluctuation at the ceiling would appear less and less natural the further it occurred from the S-curve. Even if the level of its extremity had been realized sometime in the past, if this was very long time ago, the fluctuation would seem rather unnatural. And this is the reason that chaotic fluctuations at the ceiling are to be considered as generally unnatural. They are mostly far removed from the initial rise to that level. Whether they are upward or downward they appear unnatural because a realization of this value via a natural-growth process is either impossible or it took place a very long time ago.

Ted was pleased. Once again S-curves had demonstrated solid common sense. Back home he looked again at the drawing of the S-curve drawn by someone whose hands are shaking. He felt there was a way to quantify the common-sense conclusions he had drawn in his talk earlier that day. A natural fluctuation should somehow be connected to the bell-shaped life-cycle representation of natural growth.

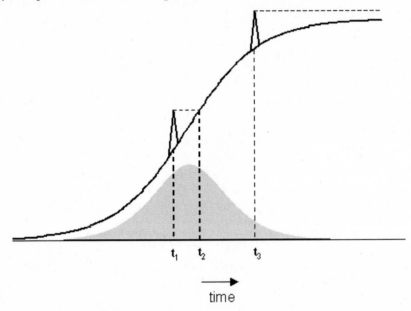

Consider a fluctuation above the S-curve at time t_1 during the curve's steep rise. The extremity of this fluctuation corresponds to the curve's level at time t_2. This fluctuation is natural because the level reached will appear in a short time in the future. But the same fluctuation at time t_3 reaches a level that has zero chance of ever being realized. Such a fluctuation can be considered as unnatural. Superimposing the bell-shaped pattern of the natural-growth life cycle on the same drawing revealed the rule Ted was looking for.

Whenever the fluctuation reaches a level that corresponds to a time within the gray bell-shaped area, the fluctuation can be considered as natural. At time t_3 the fluctuation should have much smaller amplitude to qualify for naturalness unless it was in the downward direction. For a fluctuation to be natural its extremities must correspond to points on the curve that map to the gray bell-shaped pattern, which is the natural-growth life cycle.

And as usually happened this became the subject of a yet another publication in *Technological Forecasting & Social Change*.

The intricate relationship between order and chaos proved more than intriguing; it haunted Ted. His first book was about S-curves (basically about order) and yet the phenomenon of chaos manifested itself via the very same equation (when rendered in discrete form). Douglas Hofstadter had once remarked, "It turns out that an eerie type of chaos can lurk just behind a façade of order—and yet, deep inside the chaos lurks an even eerier type of order."

Pictures of fractals endowed chaos studies with art and beauty, but Ted also discovered a certain "seductive" quality in accidents and chaotic events. For example, he was always surprised by how enjoyable it was to turn on the radio and hear a favorite piece of classical music. He could have easily reproduced the same music on his hi-fi system at home but there was more pleasure in finding it unexpectedly. As radio listening steadily loses popularity, it could be that one of its saving graces is not knowing what to expect when you turn it on.

The element of accident plays primordial importance in the formation of a couple. Accidental encounters are enchanting; they seem to be under the spell of a fairy godmother. In contrast, many highly promising matches have failed just because the meeting had been arranged and there had been no accident involved. Ted became well aware of this, so when he wanted to introduce his ex-colleague, Marsha, to his friend Tom in New York he arranged an "accidental" meeting.

Ted thought Marsha would be an ideal match for Tom. He arranged to see Marsha in Geneva before one of her trips to New York. He told her that when he was last in New York, he had borrowed Tom's golden pen and forgotten to return it. The pen was precious to Tom because it had belonged to his father. Ted asked Marsha if she wouldn't mind meeting Tom in New York and giving it to him personally.

Ted said nothing else to either Marsha or Tom. He left everything else to fate, but he did go into the trouble of buying an old-fashioned golden pen.

One year later Tom and Marsha asked Ted to be their best man!

44 – Accelerating Trends and The Singularity Myth

Working as an independent consultant proved for Ted a rewarding experience. His initial anxiety defused once he made the mental jump into becoming financially independent. It took him several months but at the end he realized that the coveted situation of not depending on employment for survival is largely a state of mind; and this is true for a surprisingly large fraction of the population in the western world. He enjoyed immensely scheduling his own time and choosing the topic of his research as he pleased.

One day his daughter showed him her cell phone. While she was going on explaining its latest features, Ted pointed out that there were even newer features that it was missing. She said she knew about them, and suddenly turned toward him in desperation, "technology is moving too fast!" she exclaimed. He thought he discerned a glimpse of terror in her glance.

He had noticed the accelerating rate of change and not only concerning technology. But his knowledge of S-curves assured him that exponential growth is encountered only during the early stages of natural-growth processes. Eventually natural growth must slow down and therefore the frenzied appearance of change in our lives could not expand forever.

A week later he traveled to America for a consulting project and among other engagements he also got in touch with his friend of old times Steve, who was now professor of cognitive and neural systems at the University of Boston. They hadn't seen each other since they worked together on the Brookhaven experiment doing their theses.

The reunion was elating. They spent hours reminiscing about old times and recounting new experiences. At one point Ted mentioned his belief that the exponential rate of change should eventually begin slowing down. Steve jumped. Apparently it was a topic he had been preoccupied with. In fact, he had tabulated what he called the twenty most significant turning points in human evolution, such as the discovery of the steam engine and the printing press, the transistor, computers, the Internet, and the sequencing of the human genome. Obviously, Steve's milestones were crowding toward recent times. In fact, he was able to demonstrate that they had followed an exponential pattern for 15 billion years!

Ted was excited. He thought this topic could yield a stimulating publication, but it was imperative to find "objective" data. Steve's twenty milestones could not stand up to the likely criticism of having a biased data set. From their experience with physics Ted and Steve knew that one needs to invest five times more effort and time into data collection than into analyzing the data and publishing the results. But Steve also had regular administrative and teaching duties to attend to, so Ted took upon himself to follow up this project.

What are the most significant turning points in history *objectively* speaking? Answers to this question can be found in compilations of most-significant-milestones lists, a favorite intellectual pastime and object of diverse academic endeavors. An example was John Brockman's book *The Greatest Inventions of the Past 2000 Years*. It was a collection of often-whimsical opinions from an elitist group of intellectuals, some of whom Ted had met among the faculty in DUXX, school-experiment in Mexico. The participants had lots of fun responding to the call and produced a list of about 75 inventions that included such "exotic" milestones as the appearance of free will, ego, and the idea of an idea.

Less witty but answering the more relevant question "What are the major events in the history of life?" are lists that can be found in the National Geographic magazine, or compiled during scientific gatherings such as symposia convened by biologists and ecologists. Still closer to the question were lists in more conventional depositories of macroscopic knowledge, such as the *Encyclopedia Britannica* and the American Museum of Natural History. They both offered compilations of "Major Events in the Universe's History." Ted found similar lists in *Scientific American* and in a number of books by scientists. To these he added Carl Sagan's celebrated *Cosmic Calendar* that matches the entire history of the Universe onto one year, pointing out dates of major events.

But Ted also used another technique, writing letters. He wrote to over sixty Nobel Prize laureates (in physics, chemistry, and medicine) asking them for what they considered to be the twenty-five or so most significant turning points in the evolution of the world. The response was very poor. He received a handful of answers and only one complete set of milestones, from biochemist Paul D. Boyer.

In any case, he distilled thirteen reliable sources of data and combined them to produce what he called the set of 28 *canonical* milestones. Indeed they coalesced in recent times. The fact that the milestones were of *top-most* importance made them of *comparable* importance. Ted then argued that the importance of each milestone was proportional to the amount of change introduced but also proportional to the distance to the next milestone (an important change carries long-term consequences.) Thus he was able to build a quantitative graph for the evolution of change that he then fitted to an S-curve.

The 28 "canonical" turning points, highlighted in bold face below, generally represent a cluster of many milestone events. The years mentioned represent an average of all of the milestones contained in each cluster and are expressed in number of years before the present time (taken as the year 2000).

1. **Big Bang** and associated processes: 15.5 billion years ago.
2. **Origin of Milky Way**, first stars: 10 billion years ago.
3. **Origin of life on Earth**, formation of the solar system and the Earth, oldest rocks: 4 billion years ago.
4. **First eukaryots**, invention of sex (by microorganisms), atmospheric oxygen, oldest photosynthetic plants, plate tectonics established: 2 billion years ago.
5. **First multicellular life** (sponges, seaweeds, protozoans): 1 billion years ago.
6. **Cambrian explosion**, invertebrates, vertebrates, plants colonize land, first trees, reptiles, insects, amphibians: 430 million years ago.
7. **First mammals**, first birds, first dinosaurs, first use of tools: 210 million years ago.
8. **First flowering plants**, oldest angiosperm fossil: 139 million years ago.
9. **Asteroid collision**, first primates, mass extinction (including dinosaurs): 65 million years ago.
10. **First humanoids**, first hominids: 28.5 million years ago.
11. **First orangutan**, origin of proconsul: 16.5 million years ago.
12. **Chimpanzees and humans diverge**, earliest hominid bipedalism: 5.1 million years ago.
13. **First stone tools**, first humans, Ice Age, Homo erectus, origin of spoken language: 2.2 million years ago.
14. **Emergence of *Homo sapiens*:** 555,000 years ago.
15. **Domestication of fire**, *Homo heidelbergensis:* 325,000 years ago.
16. **Differentiation of human DNA types**: 200,000 years ago.
17. **Emergence of "modern humans,"** earliest burial of the dead: 105,700 years ago.
18. **Rock art, protowriting**: 35,800 years ago.
19. **Invention of agriculture**: 19,200 years ago.
20. **Techniques for starting fire**, first cities: 11,000 years ago.
21. **Development of the wheel, writing**, archaic empires: 4,900 years ago.
22. **Democracy**, city-states, the Greeks, Buddha: 2,400 years ago.

23. **Zero and decimals invented**, Rome falls, Moslem conquest: 1,440 years ago.
24. **Renaissance (printing press)**, discovery of New World, the scientific method: 540 years ago.
25. **Industrial revolution (steam engine)**, political revolutions (French, USA): 225 years ago.
26. **Modern physics**, radio, electricity, automobile, airplane: 100 years ago.
27. **DNA structure described, transistor invented, nuclear energy**, World War II, Cold War, Sputnik: 50 years ago.
28. **Internet, human genome sequenced**: 5 years ago.

The S-curve indicated that year 2000 happens to be around the curve's midpoint, where the rate of growth is maximal. "We happen to be positioned at the world's prime!" wrote Ted in his article pointing out that change has never been as rapid in history and will not be as rapid in the future either. He quantified the timing of the next three milestones comparable to the 28 canonical ones as: 33, 78, and 147 years from 2000 AD.

Ted felt confident about his conclusions. Not only they were rigorously derived but they also made common sense. If one insisted that change would continue increasing exponentially, milestones should appear more and more closely together. The next milestone should be 13.4 years after the last one, the following one in another 6.3 years, the one after that in 3 years, and then again in 1.4 years, and so on. But the pattern becomes so steep that *all* future milestones are expected to appear by 2026. In other words people living at that time will have witnessed all the change that can ever take place!

Ted's article was published in *Technological Forecasting & Social Change* and later also in the larger-circulation more popular journal *The Futurist*. But it did not make the cover of *The Futurist*. The cover story of that issue was an article on the "Singularity" defined by alarmists as the approaching point in time when technological change will become so rapid that machines take over and we may enter a post-human era.

Ted contacted the BBC and suggested his article as a topic for the program *Horizon*. They thanked him for the suggestion but declined, saying that his treatment was too speculative. *Horizon* was becoming a hard-core scientific show and wanted to compete with scientific-journal publications.

Three years later Ray Kurzweil's book *The Singularity Is Near* was published and Ted, being on the advisory board of *Technological Forecasting & Social Change,* was asked to review it. It is a large book, 650 pages, and carries the accelerating rate of change to extremes. It argues in favor of double exponential trends, that is, trends that grow as an exponential of an exponential. Still, Kurzweil manages to enrobe everything in some optimism so that when machines take over sometime in the near future, humans do not become second-class citizens. Ted was annoyed by the fact that Kurzweil had relied heavily on Ted's own published articles, wrote a book that received no scientific endorsements, and yet attracted more attention and sold more copies than Ted's books. He began his review as follows:

> "This book dragged me back into a subject with which I am familiar. In fact, ten years ago I thought I was the first to have discovered it only to find out later that a whole cult with an increasing number of followers was growing around it: the "singularitarians." I took my distance from them because at the time they sounded nonscientific. I published on my own, adhering to a strictly scientific approach. But to my surprise the respected BBC television show *Horizon* found even my publications 'too speculative.' In any case, for the BBC scientists the word singularity is reserved for mathematical functions and phenomena such as the Big Bang.
>
> "Kurzweil's book constitutes a most exhaustive compilation of 'singularitarian' arguments and one of the most serious publications on the subject. And yet to me it still sounds non-scientific. Granted, the names of many renowned scientists appear prominently throughout the book, but they are generally quoted on some fundamental truth other than a direct endorsement of the so-called singularity. For example, Douglas Hofstadter is quoted to have mused that 'it could be simply an accident of fate that our brains are too weak to understand themselves.' Not exactly what Kurzweil says. Even what seems to give direct support to Kurzweil's thesis, the following quote by the celebrated information theorist John von Neumann 'the ever accelerating process of technology … gives the appearance of approaching some essential singularity' is significantly different from saying 'the singularity is near.' Neumann's comment strongly hints at an illusion whereas Kurzweil's presents a far-fetched forecast as a fact.
>
> "What I want to say is that Kurzweil and the singularitarians are indulging in some sort of para-science, which differs from real science in matters of methodology and rigor. They tend to overlook rigorous scientific practices such as focusing on natural

laws, giving precise definitions, verifying the data meticulously, and estimating the uncertainties. Below I list a number of scientific errors in Kurzweil's book. I try to rectify some of them in order to present my critique of the Singularity concept properly."

Ted then proceeded to confront the techniques, presentations, and conclusions in the book with a more scientific approach. He finally concluded:

"Scientific sloppiness is a contradiction in terms. Kurzweil and the singularitarians are more believers than they are scientists. Kurzweil recounts how he agreed with a Nobel Laureate during a meeting, but I suspect that there is no Nobel Laureate who would agree with Kurzweil's thesis. The Number One endorsement on the back cover of his book comes from Bill Gates, whose scientific credentials stop at college dropout in junior year.

"One Nobel Laureate, Paul D. Boyer, whose data Kurzweil uses when he makes his central point, has anticipated two future milestones that are very different from Kurzweil's. Boyer's first future milestone is 'Human activities devastate species and the environment,' and the second is 'Humans disappear; geological forces and evolution continue.' I have estimated that the next milestone should be around 2033. I suspect that there are many hard-core scientists who would agree with Boyer's first milestone and my time estimate.

"One could argue that Boyer is himself acting as a believer rather than a scientist, and he could be right. But Boyer did not go on to write a 650-page book on the subject. Maybe because it simply wouldn't sell!"

Following publication of Ted's review, he was contacted by the "singularitarians" inviting him to join their board of advisors in the Acceleration Studies Foundation. He would thus be contributing some of the missing scientific rigor and advancing the field toward becoming a discipline in the universities. He refused. He told them that he did not believe that machines would take over society or that there was a need to develop a discipline around how to continue to exist in accelerating change.[*] He reminded them

---

[*] In a book revisiting the Singularity concept—*The 21ˢᵗ Century Singularity and Global Futures* by A. Korotayev and D. LePoire (Eds.), Springer (2020)—Ted pointed out that three milestons predicted by the exponential trend of the singularitarians, namely in 2008, 2015, and 2018 had not shown up!

of the theorems, which say that if a machine is expected to be infallible, it cannot also be intelligent. Powerful computers are not supposed to make mistakes and therefore there can be no threatening competition to human intelligence from them. Society in any case, he concluded, has demonstrated an ability to auto-safeguard and auto-regulate itself.

He had given a similar answer to someone who after reading his first book asked him whether he was religious and whether he had faith.

"Yes, I have faith in science," Ted had answered, "which has shown me that society, like living organisms and species, is survival-prone. I would expect larger entities like the biosphere often referred to as Gaia, the solar system, galaxies, and even the whole cosmos to behave similarly."

45 – A Paradise Paradigm

Indeed there has been continuous and uninterrupted evidence for society's ability to take care of itself. Disasters and doomsdays have been prophesied time and again but all in vain. Fears of a nuclear holocaust, of running out of oil, of the HIV virus wiping out the human species, of the US deficit destroying the economy, and so on, have all proven unfounded. There is no need to invoke the existence of God or Mother Nature's superpowers to explain it. Collective human behavior *in tune* with on-going natural-growth processes sufficed each time to avert disaster by timely action often taken subconsciously. Why should it change now? Something invariant over a long period of time is easy to forecast.

In the hippie days of the 1960s there had been calls in favor of the quest for the "know-why." Ted cannot forget a cartoon in *The New Yorker* depicting two bearded gurus in a mountain cave looking at the white streak left on the blue sky by a jet airplane.

"They have much know-how, these Americans," says one to the other, "but do they have any know-why?"

The seasons metaphor Ted used in his consulting practice *naturally* placed the "why" at the beginning and at the end of the growth cycle, i.e., in "winter" season, whereas the "how" became the primary preoccupation in the "fall" season, when the growth process approached completion and the rate of growth declined.

If we imagine a large-scale S-curve depicting the evolution of human civilization from the appearance of the first cities and the techniques for starting fire to today, the emergence of science should be positioned somewhere around the 70 – 80 percent level of the curve's completion. This puts the emergence of science in the season of "fall". It also puts the gurus preoccupied with the "know-why" in *The New Yorker* cartoon either a little ahead of their time or much behind. In either case the picture is coherent!

Scientists stubbornly refuse to address the "why," leaving it to the philosophers. They even avoid making value judgments, that is, to express their opinion on whether something is good or bad. For them what people think and say is irrelevant. Only what is observable is relevant, in other words, what happens and what people do. Actions are observable.

As physics evolved over the centuries it occasionally redressed itself by going back to emphasizing the observable rather than the thinking. According to Darwinian logic, if the results are good, the procedure *must* be correct. Had another procedure given better results, it would have been chosen over the one observed.

Laws obtain their validity from their persistent existence. In the search for a purpose in life survival is a prerequisite. Laws "survive" by their nature therefore they provide a platform in the quest for exploring what life is all about.

The laws also show how God "works." Action is minimized in natural phenomena. For example, sunlight bends as it enters water, and curves around a massive object such as the Sun (as predicted by general relativity). In both cases light follows the fastest way (least action). One hundred percent efficiency is the ideal limit—inaccessible to humans—for the perfect engine. Laws reveal the direction toward becoming God-like, which enhances our chances for survival, the first of all goals.

Ted arrived at the conclusion that we are living in some form of paradise in which survival is linked to optimization and evil can only be short-lived. This is so because as Marchetti put it, the "system" behaves like an organism, with sensors, reactions, intelligence, wisdom, whims, and above all, rock-solid self-consistency. The system is one organism made out of all of us informationally linked. This anthropomorphic definition can be considered to refer to the "next step" in animal form, human beings having most of the features that cells had in primitive life forms. Our task is to clarify, quantify, discover, invent, and prepare the context for the system to act wisely. The rest will follow.

The image conforms to Gurdjieff's as-above-so-below doctrine that Ted had expounded on in one of his presentations at Chandolin. He found much evidence that at some level whatever happens is optimum just because it was allowed to happen. A rabbit population reaches the maximum number that a grass range can feed, just as car accidents per population are confined to the "optimum" number considering the car's utility. A girl playing hard-to-get optimizes the well-being among peer youngsters, and the presence of in-hospital infections permit the hospital to treat many diseases.

A city like Athens grew to over 5 million people in an optimal way but in the absence of city planning. The city developed by itself according to a simple energy-saving principle. It has a structure of many small communities defined by the common use of most basic facilities. The total person-kilometers for each one of these communities is about the same. Whenever population density increased in one community—and consequently the total person-kilometers—new facilities were created and the community split into two. The balance at any time was between paying for traveling and paying for creating new facilities. It was all done in such a way as to optimize the expenditure of energy, the biological equivalent of money.

Obviously for facilities used rarely (such as a theater, swimming pool, and park), one would be willing to walk further; these types of services were present in only one out of seven communities, but also serviced the six closest ones. This hierarchy continued upward, so that for services used even more rarely (such as a stadium, concert hall, and museum) one would have to travel even further, and more communities would share the same facilities. Interestingly, the ratio seven was maintained between hierarchical levels. Furthermore, a study identified a total of five such hierarchical levels nested like Russian dolls.

The overall structure is highly ordered and scientifically describable but does not come in conflict with individual free will. Every Athenian can buy bread at any bakery in town, but most probably he or she will buy it at the nearest one. This is the key that makes the system work. A computer study to calculate energy expenditure for different configurations of the distribution of services gave in all cases an increased amount of energy spent. The city of Athens as it developed bottom-up was naturally optimized.

We can find scientific laws behind the way things are done in nature just as we can find mathematical rules behind great works of art such as the Parthenon or the Notre Dame. In so doing we often make matters more complicated (like the mathematical description of riding a bicycle!) The usefulness of this exercise may be in part hidden and unexpected. It may not lie in the science itself but rather in finding new insights with this science, and sometimes in simply appreciating the hidden order/intelligence. Discovering that explicit laws can govern beauty eliminates a certain amount of randomness behind the origins of beauty.

We live in a paradise paradigm. As Leibniz put it in 1710:

> "If, among all the possible worlds, none had been better than the rest, then God would never have created one."

46 – Science with Street Value

While visiting CERN, Paul—a physicist from Greece and friend of Ted's from old times—came to see him at home in Geneva. He was interested to know how Ted was getting along as an independent consultant. The two physicists sat on the balcony overlooking the lake, sipping drinks, and reminiscing.

"You really meant it when you said on Greek television that you want to commercialize science. Is your next book going to be along these lines?" asked Paul.

"Commercializing science sounds right to me," replied Ted. "There is nothing dishonorable in obtaining material benefit from creative thought. Dishonor may come only with price tags because they represent an unreliable way of assigning value.

"People in academia publish for each other. Their books are difficult for non-experts in the field to read. Their articles, which appear in journals of small circulation, seem cryptic to the layperson. These people belong to oligarchic guilds and the knowledge-exchange channels remain mostly inside their peer group. A miniscule fraction of the population becomes aware of their daily discoveries.

"A more significant fraction of the population is reached by popular-science writers. These authors are either scientists or science-knowledgeable individuals, and they too write for their own kind. Popular science may claim to bring scientific knowledge to the general reader, but the reader must be scientifically minded, to say the least, before he or she can appreciate it. Yet many more people buy this literature. Books like Chaos and A Brief History of *Time* enjoyed tremendous popularity that can hardly be justified considering the fact that the complexity of content and style has probably prevented most readers from going much beyond page twenty-five. Sales of popularized-science books are artificially inflated by social vanities, by people who would like to show that they can read science, or worse by forlorn hopes of people who thought that by buying the book they acquire the knowledge contained within it. Popularized science remains by and large an intellectual activity.

"I want to go beyond popularizing science and offer the non-scientific reader books that he or she would read—not just buy—because there are pragmatic personal gains to be achieved. I have identified pieces of scientific knowledge that could be applied to everyday life and yield exploitable insights.

I'd like to extract benefits from academic research in ways that were not necessarily intended by those who produced it. Instead of waiting one hundred years or more for fundamental research to give its fruit, I am ready to propose alternate—unorthodox perhaps—ways for benefiting from some of it right now. This may be considered treason and provoke the wrath of my physicist ex-colleagues but it would be in the interest of the consumer. Did you know that in the world's oldest religion, Hinduism, commerce comes above intellect? Commerce is third while logic/intellect is seventh in importance."

Paul knew a different aspect of the primordial importance of commerce. He had just read an article by Horan, Bulte, and Shogren, which argued that the ability of *Homo sapiens* to trade gave them a crucial competitive advantage over other rival hominid species such as *Homo neanderthalensis.*

He then added, "You are not the first to look into unorthodox social applications of academic findings. I know of several individuals, mostly restless physicists, who have done similar things. Elliott Montroll, for example, was a solid-state physicist at the University or Rochester. He published a book called *Quantitative Aspects of Social Phenomena,* discussing mathematical models for social phenomena. He wrote about how competition affected traffic and population growth. He suggested that money is transferred from one individual to another in a way analogous to that in which energy is transferred from gas molecule to gas molecule by collisions. Through the transfer of goods and services everyone has some annual income. One might expect that after many transactions money will be randomly distributed, but in fact, some people end up with significantly larger incomes than others. Still, Montroll takes advantage quantitatively of some laws of thermodynamics.

"Also, Alfred Lotka—a physicist at the University of Johns Hopkins—demonstrated in the early 20th century that the growth of the railway networks in America followed the growth pattern of a sunflower seedling quantitatively. He successfully predicted the final size of the US railway network half a century before it was completed.

"In the 1960s, Derek de Solla Price at Yale developed scientometrics, that is, quantitative studies on the productivity of the scientific community. He thus created a tradition of historiography and sociology of science. He demonstrated that scientific discoveries come in clusters, and that there have been four such peak seasons in the history of the discovery of the stable chemical elements over time.

"Another one is Francis Galton, whom you may know was the grandson of Erasmus Darwin. Galton applied scientific skills on social problems as early as the mid-19th century. Among his achievements is the bringing of fingerprinting to Scotland Yard."

Ted knew them all. In fact he knew an anecdote about Galton.

"Galton wrote an essay pointing out that the number of brush strokes required for a portrait is about twenty thousand, very close to the number of hand movements that go into the knitting of a pair of socks," he said and then continued, "but more recently, besides physicists, there have been business-minded people who showed interest in drawing practical benefit from academic knowledge. Jay Forrester, who founded the System Dynamics Group at MIT, criticized the fact that social studies, physical science, biology, and other subjects are taught and generally treated as inherently different from one another while the dynamic behavior in each is based on the same underlying concepts. For example, the dynamic structure that causes a pendulum to swing is identical to the structure of forces that causes employment and inventories to fluctuate in a product-distribution system and in economic business cycles. He attributed the difference between great advances in technology and lack of progress in understanding economic and managerial systems to a failure to recognize the similarity between social systems such as families, corporations, and governments with dynamic systems such as chemical refineries and autopilots for aircraft.

"Other gurus of management consulting besides Forrester have come up with social transcriptions of scientific truths. Individuals such as Peter Drucker, Michael Hammer, Tom Peters, Peter Senge, and Shoshana Zuboff have borrowed ideas from biology, the hard sciences, and psychology. They have exploited such scientific notions as self-organizing systems, natural growth, feedback loops, chaos, and fractals to form their preachings and cast their slogans: empowerment, the learning organization, business-process reengineering, autopoiesis, evolution management, and so on. These men and women of wisdom basically behave as businesspersons motivated by profit, (at $50,000 and $60,000 per public appearance, they can probably boast the highest gross margin ever achieved by a commercial endeavor). The bits and pieces of academic knowledge they have adapted to their ends proved of high commercial value."

Ted was driving to a specific conclusion with fervor.

"It is surprising that there is no provision in advanced western society for a discipline integrator. The kind of civil servant—knowledgeable about most if not all disciplines, but specialized in none—whose job would be to stake out breakthroughs, discoveries, and other advancements with the sole purpose of producing some immediate practical benefit to society. Such knowledge agents would have to "smell out" the growth potential of a discovery without understanding its details, and arrange "marriages" involving researchers, industrialists, educators, and the department of commerce.

This vocation may require talent, but it would certainly require extensive cross-discipline education and training at university level. Finally, there should be job opportunities—state or federal—for the graduates."

Paul agreed. He empathized with Ted's enthusiastic outburst. As he prepared to leave he looked at his friend with a mixture of envy and wonder. Ted was now using science to enrich his life in ways unknown to physicists at CERN who crowded themselves in gigantic experiments waiting for years before sharing among hundreds the credit for trickles of only academic knowledge. Moreover, particle physics has saturated. It has reached the ceiling of its S-curve because smaller and smaller gains in understanding required larger and larger efforts, funds, and time. Could it be that Ted had found a more practical purpose in life?

Shortly after Paul left, Ted received another visit. Answering the doorbell he was greeted by a good-looking well-dressed young couple at his door. They looked distinguished and respectable. The man spoke in fluent Greek with no foreign accent.

"Good evening, Sir," he began. "We are seeking out compatriots in the neighborhood because we have an important message," he said and paused.

Ted was pleasantly surprised. He knew of no Greeks in his neighborhood and these two people made a good impression. But before he had a chance to ask them when and from where they came, the man continued in what resembled the recitation of a memorized text.

"We are trying to contact reputable Greeks because we are concerned about the horrible things that are happening in the world that result in sorrow, tears, and a sense of hopelessness on the part of countless numbers of people."

Ted couldn't see the point, but the man went on.

"Do you consider it progress that the bow and arrow have been replaced by machine guns, tanks, jet bombers, and nuclear missiles? Is it progress when people can travel into space but cannot live together in peace on earth? Is it progress when people are afraid to walk the streets at night, or even in the daytime in some places?" the man continued.

"Fortunately this is not the case here in Switzerland," Ted interrupted him, "but what are you trying to tell me?"

"We are Jehovah's witnesses," replied the man "and we can show you the way out from the dead end the world is getting into."

Ted was instantly turned off. "I am sorry, but I am not interested," he said and hesitated. He couldn't quite shut the door at their faces as he usually did on such occasions.

"But how can you not be interested," insisted the man. "Don't you want to feel alive and joyful. Wouldn't you like to have a purpose in life, to have something to do that will fill you with energy and emotions? That will make you capable of living fully?"

The man spoke with passion. He plucked Ted's sensitive cord. This man belonged to an Order and one of a much larger scale than the Order of Mendios.

Ted was reminded of Mihali and Aris with a certain melancholy. Their camaraderie had been shattered. Aris had distanced himself by directing all his energy to the Gurdjieff group in Athens that he had founded. Mihali had become immersed in exploiting knowledge without frontiers available in the Internet and elsewhere. The three of them had ceased to get together. Another life cycle had come to an end.

The two Jehovah's witnesses stood in front of Ted's door with serene but alert facial expressions demanding Ted's response. It was obvious that they had found their purpose in life and it was a different one than Gurdjieff's relentless pursuit of the Work or the ideas discussed in Mendios. But for Ted the question of who had the right answer was now academic. His progressive involvement with S-curves and his search for other ways science unobtrusively enters people's lives gave meaning to his life. He felt an urgency to get back to his work.

"I do want all that and more," Ted admitted, "but I have already found it. It is sitting on my desk at this very moment, and my standing here talking to you takes precious time away from my pursuing it."

Ted spoke with conviction. Suddenly there was silence. All three stood still looking at each other like warriors in a face-to-face confrontation. Ted was reminded of a familiar scene in Hollywood movies; the climactic confrontation of rival chiefs who stare at each other with meaning for a while before going their ways without saying anything.

He finally broke the silence.

"So if you'll excuse me, I need to get back," he said politely and gently closed the door behind him.

He went to his desk that was covered with books, journals, and manuscript pages. His new book was mapped on his own life experience; it began factual and informative but evolved toward more emotional and philosophical issues, all along maintaining a life-supporting line to science.

As if to substantiate what he had just said, he began reading it in loud voice savoring every word like small bites of desert:

> The attempt to assess technological inventions and scientific breakthroughs over a period of hundreds of years is unrealistic. There can be little guarantee that something good—or bad—will result from a piece of academic knowledge in the distant future.

Assessing the impact of scientific breakthroughs deep into the future is largely unfeasible.

But this is not to discourage the layperson from soliciting concrete benefits from science right away.

In the Middle Ages science wished to interpret everything within a single system, from the origin of the world to the destiny of man. That approach proved of limited use to society. The real scientific success came later at the end of the Middle Ages when science began focusing on specialized practical down-to-earth questions, which yielded direct benefit, such as how to navigate with a compass, how light traverses glass, and how water flows in a pipe.

In our times science turned again toward high-level imponderables. Toward the end of the twentieth century theoretical physicists at CERN went as far as to argue that we will soon understand all there is to be understood in theoretical physics. Theories with such flamboyant names as Grand Unification Theory (GUT) and Theory of Everything (TOE) attracted much attention. However, there followed no slowdown in research activities and there was no direct social benefit stemming from these theories.

Similarly in academic departments the volume of publications keeps swelling, but the benefits to the layperson come in trickles and only in the long run, if at all. Production of knowledge increases exponentially because each publication triggers more publications. But the public's assimilation can be linear at best. You may learn something new every day, but you could not possibly learn each day more than you learned the day before. Consequently there is an increasing gap between knowledge accumulated and knowledge put to work.

It is time to turn our attention back toward more pragmatic and specialized objectives as we did at the end of the Middle Ages. In other words, one may want to reach immediately for whatever enriching, beneficial, or simply fascinating fallout can be extracted from knowledge produced in the ivory towers of academia, wherever this is possible, even if it involves unorthodox use of the scientific results.

Acknowledgments

The author is indebted to Cesare Marchetti, who has contributed enormously to making the formulation of natural growth a general vehicle for understanding society. Many of the ideas presented by the author, such as the concept of invariants, the productivity of Mozart, and the link between entelechy and forecasting originate with Marchetti.

Of great importance have been the contributions of Mihali Yannopoulos either directly in terms of content, or indirectly by his influence as a close friend and unwitting teacher.

The 6-dimensional model of nested worlds described in Section 14 comes from Rodney Collin's book *Theory of Celestial Influence*, Robinson & Watkins, London 1973.

The Escher drawings on pages 63 and 80 have been reproduced with permission from The M.C. Escher Company; respectively:

- M.C. Escher's "Circle Limit IV" © 2008 The M.C. Escher Company-Holland. All rights reserved. www.mcescher.com
- M.C. Escher's "Cube with Magic Ribbons" © 2008 The M.C. Escher Company-Holland. All rights reserved. www.mcescher.com

Finally, the author's daughter, Thea Modis, with her enthusiasm, suggestions, and timely intervention has played a crucial role in the realization and the final shape of this book.

Bibliography

Beinhocker, Eric D. 2006. *The Origin of Wealth*. Boston: Harvard Business School Press.

Brockman, John. 2000. *The Greatest Inventions of the Past 2,000 Years*. New York: Simon & Schuster.

Browning, I. 1975, *Climate and the Affairs of Men*. Burlington, VT: Frases.

Capra, Fritjof. 1975. *The Tao of Physics*. Berkeley, California: Shambhala Publications.

Castaneda, Carlos. 1968. *The Teachings of Don Juan*. New York: Ballantine Books.

Casti, John L. 1990. *Searching for Certainty*. New York: William Morrow and Co.

Collin, Rodney. 1973. *The Theory of Celestial Influence*. London: Robinson & Watkins.

Cooke, Roger M. 1991. *Experts in Uncertainty*. Oxford: Oxford University Press.

Debecker, A. and Modis, T. 1994. *Technological Forecasting & Social Change*, 46, 153-173.

De Ropp, Robert S. 1968. *The Master Game*. New York: Dell Publishing Co.

De Sola Price, Derek J. *Little Science Big Science ... and beyond*. New York: Columbia University Press.

Feynman, Richard P., Leighton, Robert B., and Sands, Matthew. 1966. *The Feynman Lectures on Physics*. New York: Addison-Wesley Publishing Co.

Forrester, Jay. 1991. System Dynamics and the Lessons of 35 Years, in *A System-Based Approach to Policy Making*. Kenyon B. De Green (ed.). The Netherlands: Kluwer Academic.

Friedlander, Ira. 1972. *Year One Catalog: A Spiritual Directory for the New Age*. New York: Harper & Row.

Gleick, James. 1988. *Chaos*. New York: Viking.

Grübler, Arnulf. 1990. *The Rise and Fall of Infrastructures*. Heidelberg: Phusica-Verlag.

Guénon, René. 1931. *Le symbolisme de la croix*. Paris: Les Editions Véga.

Gurdjieff, G. I. 1950. *All and Everything*. London: Routledge & Kegan Paul.

Handy, Charles. 1994. *The Empty Raincoat*. London: Hutchinson.

Hawking, Stephen. 1996. *A Brief History of Time*. New York: Bantam.

Hawkings, Gerald S. 1966. *Stonehenge Decoded*. London: Souvenir Press.

Hofstadter, Douglas R. 1979. *Gödel, Escher, Bach*. New York: Penguin Books.

Horan, R. D., Bulte, E., and Shogren, J. 2005. How Trade Saved Humanity from Biological Exclusion: An Economic Theory of Neanderthal Extinction. *Journal of Economic Behavior and Organization*.

James, William. 1882. Subjective Effects of Nitrous Oxide. *Mind*, Vol. 7.

Kondratieff, N. D. 1935. The Long Wave in Economic Life. *The Review of Economic Statistics*. Vol. 17, 105-115.

Kurzweil, Ray. 2005. *The Singularity is Near*. New York: Viking.

Lotka, Alfred J. 1925. *Elements of Physical Biology*. Baltimore, MD: Williams & Wilkins Co.

Marchetti, Cesare. 1979. On 10^{12}: A Check on Earth Carrying Capacity for Man. *Energy. Vol. 4, 1107-1117*.

————. 1983. The Automobile in a System Context: The Past 80 Years and the Next 20 Years. *Technological Forecasting & Social Change*. Vol. 23, 3-23.

————. 1985. *Action Curves and Clockwork Geniuses*. Laxenburg, Austria: International Institute of Advanced System Analysis.

————. 1989. Energy Systems-the Broader Context. *Technological Forecasting & Social Change*. Vol. 14, 191-203.

————. 1994. Anthropological Invariants in Travel Behavior. *Technological Forecasting & Social Change*. Vol. 47, 75-88.

————. 1994. Millenarian Cycles in the Dynamics of the Catholic Church: A Systems Analysis. *Technological Forecasting & Social Change*. Vol. 46, 189-196.

Meadows D. H., Meadows, D. L., Randers, J., Behrens III, W. W. 1972. *The Limits to Growth*. New York: Universe Books.

Modis, T., and Debecker, A. 1992. Chaoslike States Can Be Expected Before and After Logistic Growth. *Technological Forecasting & Social Change*, 41, 111-120.

Modis, Theodore. 1992. *Predictions*. New York: Simon & Schuster.

————. 1994. Fractal Aspects of Natural Growth, *Technological Forecasting & Social Change*, 47, 63-73.

————. 1995. La santé sans la médecine, est-ce prévisible? *Médecine & Hygiène*. No 2083.

————. 1998. *Conquering Unceratinty*. New York: McGraw-Hill.

————. 2003. The Limits of Complexity and Change. *The Futurist*. May-June

————. 2006. The Singularity Myth. *Technological Forecasting & Social Change*, 73, No 2.

————. 2007. The Normal, the Natural, and the Harmonic. *Technological Forecasting & Social Change*, 74, No 3.

Monroe, Robert A. 1974. *Journeys out of the Body*. London: Corgi Books.

Montroll, Elliot W. and Badger, Wade W. 1974. *Introduction to Quantitative Aspects of Social Phenomena.* New York: Gordon and Breach Science Publishers.

Pauwels, Louis, and Bergier, Jacques. 1964. *The Morning of the Magicians.* New York: Stein and Day.

Ouspensky, P. D. 1949. *In Search of the Miraculous.* New York: Harcourt, Brace & World, Inc.

Ram Dass, Baba. 1970. *Be here now.* San Cristobal, New Mexico: Lama Foundation

Rampa, Lobsang T. 1964. *The Third Eye.* New York: Ballantine Books, Inc.

Schumpeter, J. A. 1939. *Business Cycles.* New York: McGraw-Hill.

Schuré, Éd. 1940. *Οι Μεγάλοι Μύσται.* Athens, Greece: Αργύρης Παπαζήσης.

Steiner, Rudolf. 1961. *Occult Mysteries of Antiquity and Christianity as Mystical Fact.* New York: Rudolf Steiner Publications.

Stewart, Hugh B. 1989. *Recollecting the Future.* Homewood, IL: Dow Jones-Irvin.

Vacca, Roberto. 1973. *The Coming Dark Age.* New York: Doubleday.

Virirakis, J. 1971. Population Density as the Determinant of Resident's Use of Local Centers. A Dynamic Model Based on Minimization od Energy. *Ekistics.* Vol. 187, p 386.

Volterra, V. 1931. *Leçons sur la théorie mathématique de la lutte pour la vie.* Paris: Gauthier-Villars.

Zahavi, Y., Beckman, M. J., and Golob, T. F. 1981. *The Unified Mechanism of Travel (UMOT)/Urban Interactions.* Washington, DC: U.S. Departmentof Transportation.

Zukav, Gary. 1979. *The Dancing Wu Li Masters.* New York: Bantam Books.

Index

Entries in *italics* refer to publications
Page numbers in **bold** refer to principal discussions.